ced

Pomegranate

for Nutrition, Livelihood Security and Entrepreneurship Development

The Editors

Prof R.K. Pal is currently working as the Director of the ICAR-National Research Centre on Pomegranate at Solapur, Maharashtra. After his formal education from West Bengal, he received M.Sc and Ph.D. degrees in Horticulture from IARI and joined Agricultural Research Service of ICAR. He served IARI in different capacities for over 25 years as Scientist, Professor and Head of the Division of Post Harvest Technology. He underwent advanced training in USA and attended International Course on Food Processing as FAO Fellow in the Netherlands. Based on his momentous contribution he has been conferred with the prestigious Fellowships of the NAAS and Horticultural Society of India. He is also the recipient of Dr. J.C. Anand Gold Medal of HSI. He was also appointed as National Consultant of FAO of the UN, core team member for establishment of ACARE in Myanmar and technical expert of SAARC. He has more than 100 research publications both in national and International journals, 3 books and several book chapters to his credit and filed several patents on innovative technologies that are being commercialized for Entrepreneurship and skill development.

Dr. N.V. Singh is currently Scientist at ICAR-National Research Centre on Pomegranate, Solapur. He received B.Sc. (Agri) degree from Banaras Hindu University and M.Sc. and Ph.D. in Horticulture from IARI, New Delhi. He joined Agricultural Research Service of ICAR in 2009 and has more than 100 publications to his credit. As the lead innovator, he commercialized 3 technologies on pomegranate propagation and awarded Fellowship and Young Scientist awards for his scientific contribution. He has experience of supervising more than 10 institutional and externally funded research and developmental projects. He underwent advanced training at West Virginia State University, USA.

Pomegranate

for Nutrition, Livelihood Security and Entrepreneurship Development

Editors

R.K. Pal

N.V. Singh

2017

Daya Publishing House®

A Division of

Astral International Pvt. Ltd.

New Delhi – 110 002

© 2017 AUTHORS

ISBN   9789387057210 (International Edition)

Published by   :   **Daya Publishing House®**
A Division of
Astral International Pvt. Ltd.
– ISO 9001:2015 Certified Company –
4736/23, Ansari Road, Darya Ganj
New Delhi-110 002
Ph. 011-43549197, 23278134
E-mail: info@astralint.com
Website: www.astralint.com

Dr. K.L CHADHA
Ph.D (Hort.), D.Sc. (Honoris causa),
FNAAS, FHSI, FISHS
(Padma Shri Awardee)
President
The Horticultural Society of India
F-1, Societies Block, NASC Complex,
New Delhi-110 012
Tel.: (011-25842127
E-mail.:-hsi42@rediffmail.com

Foreword

Formerly
National Professor (Hort.) ICAR,
New Delhi
Deputy Director General (Hort.) ICAR,
New Delhi
Horticultural Commissioner,
Min. of Agri., GoI, New Delhi
Executive Director, NHB,
Min. of Agri., GoI, New Delhi
Director, IIHR (ICAR), Bangalore
Project Coordinator (Fruits) I CAR,
Lucknow

Pomegranate (*Punica granatum* L.) cultivation today is a highly lucrative and remunerative agri-business in India. The alluring monetary returns per unit area from this crop have resulted into remarkable increase in area by about two folds and production by about three folds within a short span of five years. Its versatile adaptability to a wide range of climatic conditions, hardy nature, low water requirement, good response to high tech-horticultural practices, high yield, magical therapeutic values and increasing demand for fresh consumption, processing and export has led to its immense popularity in recent times. Pomegranate acquires centre stage in the livelihood security of farmers in dry Deccan Plateau region of India owing to its less natural resource demanding nature. Latest research and development initiatives suggest huge potential of pomegranate in processing with array of nutritionally loaded value added products including juice, ready to serve drinks, wine, seed oil etc. Pomegranate and its high value processed products can play a vital role in doubling the farmers' income by 2022, if supported ably by sound scientific cultivation practices and effective value chain.

A holistic literature to create awareness about immense potentiality in pomegranate and supplement the much needed scientific information on pomegranate improvement,production, integrated disease and pest management, processing and entrepreneurship opportunities, is the need of the hour. The present book entitled *"Pomegranate for nutrition, livelihood Security and Entrepreneurship Development"* is an excellent compilation of latest scientific information on pomegranate which will be equally useful to researchers, academia, entrepreneurs and progressive growers. The meticulously designed content and scientific information by experts on pomegranate make this book a complete package and a must read literature for all the stake holders of pomegranate.

I congratulate and also express my appreciation for the sincere efforts made by Dr. R.K. Pal and Dr. N.V. Singh in bringing out this well compiled informative publication.

(K.L.Chadha)

Preface

Pomegranate (*Punica granatum* L.) is one of the important fruit crops in arid and semi-arid regions. It is cultivated extensively in Turkey, Tunisia, Israel, Egypt, Spain and Morocco, Iran, Afghanistan, India and to some extent in the U.S. (California), China, Japan and Russia. Pomegranate is the only crop that gives the highest return on investment in the dry land agro-eco system. Maharashtra state is considered as the 'pomegranate basket of India' which contributes more than 70% of the total area under pomegranate cultivation followed by Karnataka, Gujarat and Andhra Pradesh. Its area is also rapidly increasing in Himachal Pradesh, Rajasthan and Madhya Pradesh. India produced 2.198 million tonnes of Pomegranate from an area of 0.193 million ha as per the revised estimate of NHB 2015-16. Although India is the largest producer of pomegranate in the world, its national average productivity is 11.38 t/ha which is far below the productivity of Turkey (27.25 t/ha), Spain (20.00 t/ha), USA (16.7 t/ha). The main reasons of low productivity are lack of improved varieties, good agricultural practices, prevalence of hostile agro-climatic condition and severe biotic and abiotic threats in the growing regions.

Pomegranate fruit is a rich source of riboflavin. Pomegranate arils provide 12 and 16 per cent of daily dietary requirement of vitamin-C and vitamin-K, respectively, per 100g serving. It is also a rich source of micronutrients *viz*. Iron and Zinc. Ellagitannins and flavonoids are most important bio-active components present in pomegranate that have been reported highly beneficial to combat several human ailments. The edible seeds of pomegranate contain 15-25 per cent oil which is a rich source of Conjugated Linoleic Acids (CLA) *viz.*, punicic acid, palmitic acid, stearic acid, oleic acid and linoleic acid that have been reported to delay the aging animal cells, reduce the atherosclerosis and thereby help in preventing various heart ailments. Pomegranate seed oil also has been found to possess anti-inflamatory and anti-tumour properties with possibility of anti-cancer effect. Pomegranate seed oil has great potentiality in developing several pharmaceutical and cosmetic products. Rind of fruit, bark of stem and root of pomegranate contain more than 28 per cent gallotannic acid that could be extensively used as natural yellow dye for textile industries as bio-colour. However, commercial exploitation on value chain of pomegranate in India is either scanty or nil. Pomegranate fruits have vast demand both in national and international market. India exports around 35,000 tonnes of fresh fruits per annum.

In view of the great potentiality of pomegranate based farming system in dry land agriculture publication of the book entitled '**Pomegranate for Nutrition, livelihood Security and Entrepreneurship Development**' is the very timely and need of the hour. I appreciate the efforts of all the contributors and especially Dr. Jyotsna Sharma, Dr. Ram Chandra, Dr. K.D. Babu, Dr. Ashis Maity, Dr. D.T. Meshram, Dr. Nilesh Gaikwad, and Sh. Mallikarjun of ICAR-NRC on Pomegranate for their significant contribution in compilation of the book in its present form. I am equally thankful to Dr. N.V. Singh for his painstaking effort in jointly editing the book. I am also thankful to M/S Astral Publishing House, New Delhi for kindly agreeing to bring out this publication on the occasion of the 2nd National Seminar -Cum-Farmers Fair on 'Pomegranate for Health, Growth and Prosperity' at Solapur during April 28-30, 2017 jointly Organized by the Society for Advancement of Research on Pomegranate (SARP) and ICAR-NRCP. I hope this book will be of immense help to many researchers, students and stakeholders directly or indirectly associated with pomegranate based farming system and entrepreneurship development.

(R.K. Pal)
President, SARP

Contents

1

Developing Pomegranate Varieties for Livelihood Security by Conventional Approach

S.H. Jalikop

ICAR-Indian Institute of Horticultural Research, Bengaluru-560 089, Karnataka
E-mail : jalikop@iihr.ernet.in

Introduction

Pomegranate (*Punica granatum* L.) is native to hot dry regions of Afghanistan and Baluchistan (De Candolle, 1967), but with time it has diffused and got adapted to a wide range of climatic conditions. It produces fruits in tropical, sub-tropical (Trapaidze and Abuladze, 1989), temperate (Purohit ,1982; Levin ,1995) regions and in hilly areas up to 1800 m altitude (Sharma and Sharma, 1990). The unique plasticity of this fruit crop is evident from the threshold limits it exhibits for high (44°C) and low (-12°C) temperatures (Westwood, 1978). The plant habit, fruiting and flowering physiology are, however, altered with the habitat (Pareek and Godara, 1993). In India this ancient fruit was grown for long as a backyard tree but in the recent times it has emerged as a commercially important fruit.

Indian farming community is dominated by small and marginal farmers with less than 2 ha land holdings (Gautam *et al.*, 2007). Resources for farming at the disposal of these marginal (<1ha) and small (<2ha) farmers are too little. They are neither technologically nor financially well placed. According to Chand *et al.* (2011), the share of small and marginal farmers in land holdings is as high as 83 per cent. About 98 million out of total 120 million farm holdings are small and marginal farmers (Dev, 2012). Nevertheless the role of small and marginal farmers in development and poverty reduction is well known (Lipton, 2006). In terms of production, these farmers also make large contribution to the production of high value crops. They contribute around 55% to the total production of fruits (Birthal, 2011). Thus, small and marginal farmers contribute to both diversification and food security. It is an irony despite this contribution, these farmers comprise almost three fifth of the nations hungry and poor (Gautam *et al.*, 2007).

Pomegranate is an ideal fruit crop for sustainability of small holdings and it comes well in the hostile ago-climate of arid and semi-arid regions. Hence availability of hardy drought tolerant pomegranate varieties with acceptable fruit yield and quality should help in providing livelihood security to the resource poor small farmers as 90 per cent of them depend on rain (Gautam *et al.*, 2007). Rainfed regions are generally characterized by lower economic returns from agriculture and a high concentration of small and marginal farmers (Ganguly, 2011). Unlike large farmers small farmers can ill-afford expensive inputs required for a good crop like planting material, labour, fertilizers, pesticides or opening of bore wells. Cultivation of high yielding varieties require much greater technical expertise, working capital, and marketing network than the traditional crops. Thus several small farmers who adopted cultivation of high yielding varieties of cereals during green revolution were not successful as these varieties demanded intensive cultivation. Besides the high yielding crop varieties narrowed the genetic base almost dangerously, depleting the rich genetic wealth of the farmers' fields and threatening the future of plant breeding and the creation of new crop varieties.

Indian pomegranate industry hinges mainly on a single variety, 'Bhagwa'. Area under this variety has increased significantly in the recent times and has marginalised long standing variety 'Ganesh'. 'Bhagwa' produces a heavy crop of high quality fruits suitable for export and offers attractive returns. There is good demand in international markets for fruits of 'Bhagwa' while 'Ganesh' is still a preferred choice in domestic markets (Ganguly, 2011). 'Bhagwa' variety requires intensive cultivation with high inputs, which often is difficult for the small and marginal farmers to afford. Thus this variety is mainly in the hands of progressive rich farmers. Predominance of a single variety and over dependence for pomegranate production on it poses great risk. The sudden outbreak of a devastating pest or disease might threaten the very existence of pomegranate industry in India. Hence there is an urgent need to develop new varieties that suit both resource rich farmers and resource poor small farmers and help to ameliorate the livelihood security of small farmers.

Global Varietal Wealth

In different pomegranate growing areas of the world a good number of varieties have been identified by growers and recently by breeders. The Turkmenistan Experimental Station of Genetic Plant Resources, Turkmenistan Academy of Agricultural Sciences, Garrygala, has a large collection of 1157 accessions (Levin, 1995), Iran has about 760 germplasm and China grows 238 cultivars in different provinces (Feng, 2006). Though such large diversity is available across the world very little germplasm has been exploited in terms of breeding work. A review of varietal status shows that the traditional varieties in different countries are being replaced by modern ones.

Important Varieties: In addition to seedling selection other breed approaches like hybridisation, mutation and polyploidy work in pomegranate have resulted in number of new genotypes/varieties. Some important varieties developed are: 'Hongmanaozi', 'Taihanghong', 'Yushiliu 4', 'Vurgun', 'Aleko', 'Mengliaihong', 'Zaoxuan 018', 'Zaoxuan 027', 'Taishan Dahongshiliou', 'Huashu', 'Dahongshiliu', '87-Qing 7', 'Linxuan 8', 'Lintong 14', 'Baiyushizi', 'Qingpiruanzi', 'Tiepitian' (China); 'Shhani-Yonay' (Israel); 'Azerbaijan', 'Nasimi' (Azerbaijan); 'Desertnyi' (Soviet Union); 'ME14', 'ME 15' (Spain). Currently in India pomegranate cultivation is virtually dominated by 'Bhagwa' with several parallel names, and 'Ganesh' and its variants, of course the former variety by far dominates the later. However a couple of other varieties of local interest are being grown on a very limited scale.

Ornamental Varieties: Certain *Punica* forms have ornamental value. There are mainly two ornamental types: 'Double Flower' and 'Dwarf' or 'Nana' types. 'Double Flower', as the name implies, produces double flowers wherein numerous stamens are modified into petals; as a result many of them do not set fruits, but because they have large attractive flowers they have ornamental significance. The functionally sterile flowers, however, yield fruits when manually pollinated. Another type of pomegranate which has exclusive ornamental value is Nana, whose adult plant is a miniature pomegranate like *bonsai*. It has small leaves, grows to a height of 50-70 cm, and bears small flowers and fruits (Nath and Randhawa, 1959). Attempts have been made to exploit 'Nana' in breeding for moisture stress tolerant varieties, bacterial blight resistant varieties and to develop yellow miniature pomegranate (Jalikop *et al.*, 2003).

Wild Forms: Wild *Punica* germplasm is useful in breeding programmes as they carry valuable genes not available sometimes in the cultivated forms. Pomegranate grows wild in Transcaucasia and in Asia Minor. Some parts of the Mediterranean area are also considered as native lands of pomegranate. A wild pomegranate, *'Daru'*, is very common and gregarious in gravel and boulder deposits of dry ravines in the outer Himalayas. These hardy deciduous seedling trees growing over long periods of time possess better climatic adaptability and resistance to pests and diseases (Sharma and Sharma, 1990) owing to natural selection. Dried arils of *'Daru'* are used in the preparation of *anardana* (Pruthi and Saxena, 1984). In order to introgress bacterial blight resistance of *'Daru'* Jalikop *et al.*, (2005, 2006) hybridised it with the cultivated varieties.

Useful Donors: Holland *et al.* (2009), reviewed in detail several important pomegranate varieties available in the world. Some of these varieties can serve as useful parents while breeding for: large fruit, very juicy, sweet, pink arils – 'Ganesh', 'Teipitian'; large fruit size – 'Bopi', 'Teipitian'; red skin, pink arils – 'Duan- zhihong'; earliness – '87-Qing 7', 'Alack'; sweet- sour, green skin – 'Dabaitian', 'Heyinruanzi'; resistance to fruit splitting – 'Apsheronskii Krasnyi', 'Frantsis', 'Kyrmyz Kabukh'; high juice content – 'Slunar', 'Pirosmani', 'Vedzisuri'; attractive thick skin, late maturing –'Mridula', 'Bhagwa'; thin skin, early maturing – 'Ruby'; 'Muskat', Dholka'; thick red rind, red arils, late ripening – 'Malas-e- saveh', 'Rabab-e-Neyriz', 'Malas-e-yazdi', 'Naderi-e-bud- rood'; vitamin C – 'Siyah'; yellow skin – 'Lefan', 'Kabul Yellow'; red skin, red arils, hard seeds, sweet sour – 'Janarnar'; red glossy skin, sweet sour – 'Wonderful' ('P.G.101- 2'); sweet, pink arils – 'Rosh Hapered', 'Malisi'; sweet fruits with soft seeds, red arils – 'Mollar de Elche' and its selections 'ME1', 'ME5', 'ME6', 'ME14'; red colour, yield – 'Kamel', 'Akko'; high yield – 'ME15', 'ME16', and 'ME17', 'Agria de Blanca', 'Albar de Bianca', 'Borde de Albatera', and its selection 'BA1', 'Borde de Blanka', 'Casta del Reino de Ojos', 'CRO1', 'Mollar de Albatera Mollar de Orichula', 'Lnxuan-8', 'Hicaznar'; resistance to moisture stress – 'Taishan Dahongshiliou', 'Nana'; adaptation to sandy alkaline soils – 'Yushiliu 1', 'Yushiliu 2'; bacterial blight tolerant – 'Daru'.

Breeding Objectives

Breeding objectives differ from region to region (Holland *et al.*, 2009) and also whether the fruit is used for fresh or processing purpose. Fruits having bold, juicy, pink or red arils with small, soft seeds are preferred though likings of consumers are not always uniform. Relatively high juice content may be more desirable than large fruit size. Since pomegranate has become an economically important fruit crop and is being exported breeders should consider criteria of international markets like fruit size, fruit shape, skin colour, aril colour, easily separable bold arils, seed size, sugar acid blend and beneficial nutrients in the juice, besides ripening time, shelf life, good post-harvest quality, and handling of fruits during transport are also desirable characters. Additionally screening of vast *Punica* germplasm should receive due attention for several of the documented medicinal properties of pomegranate as genetic variation is bound to exist for the active chemical compounds. In recent times breeding for physiological disorders like fruit cracking and aril browning or internal break down of arils (Jalikop *et al.*, 2005, 2006, 2010) are also being considered. As regards plant habit, genotypes having an evenly spreading canopy with strong branches free from thorns and drought hardy are ideal. Sometimes dwarf plant habit is a desirable characteristics for mechanical and easy hand harvesting (Glozer and Ferguson, 2008). Region-specific objectives for example, in hot dry regions where water is scarce, high temperature and moisture stress tolerance, and in India where bacterial blight is rampant, blight resistance should receive due importance in breeding programmes (Jalikop *et al.*, 2005). In temperate regions breeding for cold hardiness is important.

Developing Varieties for Resource Poor Small and Resource Rich Large Farmers

The risk taking capacity of small and marginal farmers with limited resources is low. These farmers with little capital select varieties which require minimum inputs while large rich farmers prefer high input varieties which give lucrative returns. The commercially oriented resource rich farmers often select such varieties that produce fruits suited for international markets or for supermarkets as it happens to be most remunerative. Breeding varieties for resource rich farmers is more specific and exacting, hence more challenging. Small and marginal farmers prefer to grow hardy varieties that provide livelihood security. The resource poor farmers often use their produce for selling in local / domestic markets or self-consumption or supply to processing industry. In this paper developing varieties that meet these two different goals that is varieties that suit to resource poor and rich farmers is discussed.

Irrespective of whether a variety is developed for small and marginal or large farmer breeder has to give due thrust on certain traits like high yield potential, tree vigour, earliness and bold juicy arils. But in respect of some traits like fruit colour, fruit shape, seed mellowness, aril colour and acidity breeder can have some flexibility while breeding varieties for small and marginal farmers. This is because small and marginal farmers mostly supply produce to processing industry or indigenous fresh market. Soft or semi soft seeded types are used for fresh consumption while hard seeded ones are supplied for juice or wine making industry. Hard seeded types having very high level of acidity are suited for *anardana* preparation. Further, there is some scope for altering acidity, TSS and juice colour during processing the fruits for juice or wine. While reasonably good fruit size, shape and colour are preferred for indigenous markets, wide range of fruit size, shape and colour are acceptable when produce is targeted for processing. Likewise while selecting genotypes rich in medicinal properties pomologically valuable traits become less important.

This relaxation in selection for many of the traits offers better opportunity in pomegranate breeding programme. This should allow the breeder to select otherwise exceptionally superior recombinants (possessing traits like resistance to drought or bacterial blight) with acceptable yield and fruit qualities. Such selections should be further evaluated for their suitability for releasing to small and marginal farmers. Therefore, in developing varieties suited for small and marginal farmers, the breeder has to practice flexible selection for certain traits (depending on purpose for which a variety is being bred) and follow a rigorous selection for other desirable traits. Ultimate objective is to breed low-maintenance cost-effective varieties for small and marginal farmers. The evaluation and selection of genotypes should involve assigning of appropriate weights for various traits by taking into account the criterion for which a variety is being developed, and ranking them based on weighted average. This should enable to judge correctly and isolate right genotypes for a specific purpose.

Breeding Strategies

Pomegranate, being often-cross pollinated crop (Jalikop and Kumar, 1990), is genetically heterozygous and adequate variation for several plant and fruit traits is generated in natural seed originated populations. Most genetic combinations that breeders may look for are probably disseminated in variety populations and wild ecotypes (Holland *et al.*, 2009). Several cultivars grown today are the result of human selection from naturally occurring intra-species variation. Very few varieties have been developed by systematic breeding programmes despite great opportunities exist for evolving varieties to meet various objectives. According to Mars (2000), pomegranate culture is still faced with many problems and methods must be developed for cultivar identification and improvement, and genetic resource management. Since there is increasing worldwide demand for this fruit for its superior pharmacological and therapeutic properties, there is a need to initiate and intensify well planned deliberate breeding programmes to meet the demands of local and international consumers, processors, growers and exporters. Built in genetic resistance to pests and diseases makes the cultivation economical and environmental friendly that will be boon for both small and large farmers. Breeding new varieties may be achieved by conventional approaches like seedling selection, developing inbred line, hybridization followed by selection or through mutation. However, mutation breeding is only a 'hit and miss' approach hence the probability of getting the targeted genotype is very low.

Seedling Selection: The presence of numerous seeds in pomegranate fruit coupled with good seed germination besides growers sometimes preferring to cultivate seedlings has made large pomegranate seedling populations available across the globe. Seed originated plants often exhibit variation which makes the selection of plants for traits of interest possible. Hence, seedling selection assumes great significance in pomegranate breeding (Jalikop and Kumar, 1998). Even professional breeders find seedling selection quite promising in developing new pomegranate varieties. The probability of getting superior recombinants is proportional to the population size used for making the selection. Selection for fruit quality can be accomplished in first year of production, although fruit weight may not be typical of when the plant bears a full crop. Fruits are often larger when production is low in young plants. Based on the phenotype once a desirable plant is located it must be propagated vegetatively (either by air layering or cuttings) and its superiority should be verified in a replicated trial in relation to a ruling variety.

In Israel, early ripening 'Emek' was selected from seedlings raised from open pollinated seeds. In Tajikstan, during 1958-1961 Rozanov (1972), singled out seedling varieties 'Apsheronsk' and 'Melesi' for flavour and yield. Chinese workers Wang *et al.* (2006) selected a chance seedling 'Zaoxuan 018' from 'Dahongpao' and 'Zaoxuan 027' was selected from 'Daqinpi'. Most popular soft-seeded Indian cultivar 'Ganesh' is a selection made from the seedling population of hard-seeded 'Alandi' by Cheema and subsequently 'G-137' was located from the open-pollinated progeny of 'Ganesh'.

Inbred Line Varieties: Pomegranate can withstand certain amount of inbreeding (Jalikop and Kumar, 1993). Selfing should preferably be initiated in seedling varieties and desirable progeny should be selected in each selfed population. The traits that matter most like seed mellowness, aril colour, TSS, fruit shape and colour should be fixed by selfing and selecting over generations. Sib mating between selected siblings can be adopted at times if inbreeding depression is noticed in any generation. Once pomologically important traits are found genetically stable which should normally happen by S4 or S5 generation, several phenotypically similar lines can be mixed (akin to mass selection) and tested for their overall performance before releasing for cultivation. Subsequent propagation of the new variety should be done using the selfed seeds.

A variety developed by this method consist of a mixture of seed propagated lines. The tap root system of seedlings in contrast air-layered or cutting-propagated or tissue culture plants having adventitious roots exploit soil moisture better besides providing a good anchorage for the plant. Inbred line varieties will be phenotypically as well as genetically dissimilar but for few selected traits. Such a varietal population is expected to exhibit low incidence of diseases and pests as compared to vegetatively propagated genetically identical population. However the fruits harvested from inbred line varieties may lack complete uniformity but the produce could find place in domestic market or processing industry or home consumption. Inbred line varieties allow farmer to raise his own planting material from the selfed seeds. Hence inbred line varieties appear to hold great potential in meeting the needs of small resource poor farmers who mostly grow the crop under rain fed condition.

Hybridization: Hybridization is a straightforward conventional approach of combining different traits available in the germplasm. The desirable hybrid progeny is selected in the F1 or in a later generation, either it can result in a new variety or the selected genotype can serve as a breeding line in future breeding programmes. Hybridization and development of a hybrid population is relatively easy in pomegranate as the flowers are big, pollen in plenty is readily available and seed germination is good; besides, many of the varieties are easily crossable (Karale and Desai, 1999; Jayesh and Kumar, 2004) and producing good percent of fruits. However when wild non-cultivated types like *'Daru'* are involved in the breeding work one has to go for repeated backcrosses by making selection in each generation in order to eliminate several undesirable traits of the wild type. Thus, developing pomegranate varieties by backcross breeding is time consuming.

Selection of Parents: Selection of appropriate parents is critical as success of new varieties developed by hybridization hinges on the parents used in the initial crossing programme. There are several varieties of pomegranate both adapted to tropical and sub-temperate climate offering a wide choice of parental combination in the hybridization programme. Many economically important traits are disseminated in world pomegranate varieties. According to breeding objectives appropriate donor parents can be located from the vast collection. Jalikop and Kumar (1998), suggested using soft-seeded cultivars as parents while breeding for high juice cultivars due to their significantly higher content of juice. Breeder has to be discreet in selection of parents as growing large hybrid population requires adequate land, labour and other resources additionally their evaluation involves time, effort and money.

Progeny Selection: As pomegranate is heterozygous segregation takes place following hybridization in the F_1 progeny itself and selection for the desired genotypes can be practiced in this generation. Jalikop *et al.* (2006), observed the occurrence of more recombinants in F_2 than in F1 hence, breeder can also look for new recombinants in F_2. Use of molecular markers and morphological markers, if available, significantly reduces the progeny population to be field planted for evaluation. Heterosis for fruit characters was reported by Karale and Desai (2000) over the mid, better and top-parent. The percentage of heterosis over the mid-parental value in the desirable direction was maximum for juice weight (86.44%), followed by aril weight (71.20%), fruit weight (65.1%), rind thickness (48.92%).

Once superior selections are made, they have to be multiplied clonally (so that the genetic integrity of the selected progeny is not lost), and tested against the ruling variety in a replicated trial before identifying for release. Multilocation evaluation is preferred before recommending for commercial cultivation. Development of varieties by hybridization and selection has been reported from India, China, Turkmenistan and Azerbaijan. A gene for red aril colour (RR) was transferred by Jalikop and Kumar (2000) from a Russian temperate variety and developed a tropical multiple hybrid variety, 'Ruby'. This implies scope for exchange of mutually beneficial genes from tropical and temperate pomegranates. Hybridization between cultivated pomegranate varieties and the wild type, *'Daru'* and the ornamental variety, 'Nana' was attempted by Jalikop *et al.* (2005, 2006), in order to transfer resistance of *'Daru'* and 'Nana' for bacterial blight disease prevalent in India. When non-cultivated types are involved in breeding, in order to eliminate several undesirable traits repeated backcrosses are necessary with the selected progeny in each generation. No inter-specific hybridization has been reported with the other species of the *Punica* genus, *P. protopunica*.

Mutation Breeding: Mutation breeding, though useful in creating new variation needs a large population to be raised to locate a desirable type. Mutations occur in a random fashion and are generally recessive and deleterious in nature. Many mutations usually do not have economic significance which makes this technique less attractive. Possibly, this is why limited experiments have been attempted in pomegranate with little success. Akhund-zade *et al.* (1977), gamma-irradiated seeds and cuttings with 1-40 kR and selected types which exceeded the initial material in fruit yield, size and quality. They also observed varieties with sweet fruit were more sensitive to radiation than those with acid-sweet or acid fruit. Levin (1990) obtained forms with good fruit and juice quality and good keeping quality in the mutant seedlings developed by treatment with gamma rays at 10-20 kR. Promising mutants were also bred by N,N-dimethyl- N-nitrosourea treatment (Levin, 1990). Akhund- zade (1981), studied gamma irradiation effect on cuttings, seeds and pollen and noted a wide range of variability at 5-10 kR doses. Recently in China some varieties like 'Hongma-naozi', 'Taihanghong' and 'Hongyushizi' were isolated as natural mutants (bud sports) from thee different varieties.

Conclusion

Conventional breeding techniques hold immense prospects in developing desirable varieties of pomegranate using large natural genetic variation accumulated over generations and numerous diverse varieties spread over in several countries. Very little deliberate effort has been made in genetic improvement of this fruit. More intensive and comprehensive breeding work with specific objectives is required. It should comprise exhaustive hybridization involving the world varieties, evaluation of a large number of progenies and meticulous selection and testing. New varieties of interest can also be recognized for a given region by simple introduction from foreign countries or by making direct selection from available natural *Punica* gene pool. Scientists should resort to mutation breeding when trait of interest is not present in the natural gene pool. Better understanding of genetics of pomegranate will guide in designing breeding programmes. However, at present available knowledge of inheritance of various fruit and plant characteristics is meager. The requirements of pomegranate fruits especially for international market include good shelf life, easily extractable, vividly coloured bold arils containing small seed or no seed with good eye appeal. Additionally fruits should have good sugar:acid blend and nutraceutically more valuable. Future research work should also focus on breeding for biotic and abiotic stresses.

In India more than 80 per cent of farmers are small and marginal farmers with less than 2 ha. holdings and 90 per cent of them depend on rain. Unlike large farmers small and marginal farmers who often face livelihood security can ill-afford expensive inputs like planting material, fertilizers and pesticides. Further small resource poor farmers mostly sell their produce in domestic markets or use for self-consumption or supply to processing industries. Small farmers look for hardy low-input but cost-effective pomegranate varieties and there is an urgent need to develop such varieties.

Breeding new varieties for large and small farmers may be achieved by conventional approaches like seedling selection, developing inbred line, hybridization followed by selection or through mutation. While availability of suitable pomegranate varieties empower small farmers and ameliorate their living standards, the big monocultural farms are simply not going to disappear.

2

Crop Improvement Strategies in Pomegranate (*Punica granatum* L.)

K. Dhinesh Babu[1], Ram Chandra[1], N.V. Singh[1], J. Sharma[1], Prativa Sahu[1], R.K. Pal[1] and B.N.S. Murthy[2]
[1]ICAR-National Research Centre on Pomegranate, Solapur-413 255, Maharashtra
[2]ICAR-Indian Institute of Horticultural Research, Bengaluru-560 089, Karnataka

Introduction

Pomegranate (*Punica granatum* L.) is an important fruit crop of arid and semiarid regions of the world. The cultivation of pomegranate by mankind as a fruit crop dates back to antiquity. The usage of pomegranate is deeply embedded in human history with references in many ancient cultures for its use in food and medicine. It is one of the oldest known edible fruits and is associated with ancient civilizations of the Middle East.

The pomegranate (*Punica granatum* L.) is believed to be originated from Iran (Primary centre of origin). Besides, it is widely prevalent in Afghanistan, Pakistan and India, the Secondary Centres of Origin (De Candolle, 1967). From its origin in the area now occupied by Iran and Afghanistan, the pomegranate cultivation had spread to India, China and Mediterranean countries viz., Turkey, Egypt, Tunisia, Morocco, and Spain. Spanish missionaries brought the pomegranate to America in the 1500s (Hodgson, 1917; La Rue, 1980).

Improvement of pomegranate becomes imperative for development of varieties suitable for table purpose, varieties suitable for processing purpose, varieties with biotic and abiotic resistance. The various breeding methods for improvement of pomegranate include selection, hybridization, mutation, *etc.*

Cytology

The genus '*Punica*' belongs to the family 'Lythraceae' (Sub-family: Punicoideae) and has two species. *Punica protopunica* found wild in Socotra island and the cultivated *P. granatum* (Anon., 1952). *Punica granatum* has 2n=16,18 chromosomes. The number of chromosomes in somatic complements of Dholka, Ganesh, Kandhari, Muscat White and Patiala was found to be 2n=16, while the variety Double Flower had 2n=18 (Nath and Randhawa, 1959). The chromosome number in Vellodu and Kashmiri varieties was found to be 2n=18 (Raman *et al.*, 1963).

Floral Biology

Floral biology of pomegranate has been reported by many workers (Nath and Randhawa, 1959; Nalawadi *et al.*, 1973; Singh, 1977, Bavale, 1978; Josan *et al.*, 1979; Game, 1987). The inflorescence is cyme which includes staminate, intermediate and hermaphrodite flowers. Both self and cross pollination were recorded in pomegranate. The fruit is a modified berry developing from the inferior ovary.

Germplasm Collection

The genus '*Punica*' has two species viz., *Punica granatum* (cultivated pomegranate) and *Punica protopunica* (wild pomegranate). The cultivated pomegranate *P. granatum* is divided into two sub-species viz., *P. granatum* subsp. *chlorocarpa* and *P. granatum* subsp. *porphyrocarpa*. The wild pomegranate (*P. protopunica)* is found in Socotra Island. Pomegranate grows as wild in Syria, Afghanistan, Central Asia and India (Saxena *et al.,* 1987). In India, wild pomegranate growing in Himachal Pradesh, Uttaranchal and Jammu & Kashmir is known as '*Daru*'. Chandra *et al* (2011) reported on the germplasm collection available in the field gene banks of India (Table 2.1).

Table 2.1: Status of Pomegranate Germplasm in India

S. No.	Collection centre	State	Number of collections
1.	CCS Haryana Agricultural University, Bawal	Haryana	09
2.	Rajasthan Agricultural University, Jobner	Rajasthan	09
3.	Punjab Agricultural University, Abohar	Punjab	19
4.	Tamil Nadu Agricultural University, Aruppukottai	Tamil Nadu	24
5.	Acharya N.G. Ranga Agricultural University, Anantpur	Andhra Pradesh	29
6.	ICAR-Central Arid Zone Research Institute (ICAR), Jodhpur	Rajasthan	34

Contd...

S. No.	Collection centre	State	Number of collections
7.	S.K. Nagar Agricultural University, S.K. Nagar	Gujarat	52
8.	Mahatma Phule KrishiVidhyapeeth, Rahuri	Maharashtra	52
9.	ICAR-Indian Institute of Horticulture Research, Bengaluru	Karnataka	64
10.	ICAR-Central Institute of Arid Horticulture (ICAR), Bikaner	Rajasthan	152
11.	ICAR-National Research Centre on Pomegranate (ICAR), Solapur (*- imported from California, USA)	Maharashtra	177+168*

Breeding Objectives

The breeding objectives may vary according to the need and situation.

- To develop table purpose varieties with soft seeds, bold red arils, attractive red rind, better yield and quality
- To develop processing purpose varieties with high acidity, better yield and quality (suitable for *anardana* making)
- To develop varieties tolerant/resistant to biotic stresses (bacterial blight, wilt, fruit borer, *etc.*)
- To develop varieties tolerant to abiotic stresses (fruit cracking, soil salinity, *etc.*). To develop dwarf statured, thornless varieties suitable for ultra high density planting

Improvement Methods

The different methods of crop improvement in pomegranate includes Introduction, selection, hybridization, mutation, *etc.*

Introduction : Some of the important pomegranate cultivars with dark red arils introduced into India were as follows.

S.No.	Varieties	Country
1.	Wonderful	USA
2.	A.Males, A. Be Hastah, A. A lah, A. Agha, Mohammad Ali, A Post SephidSirin	Iran
3.	Ranninj G 1-8-23, Rannyij G 1-3-34, J G 1-8-7	Russia
4.	Gulesha, Gulesha Red, Gulesha Rose Pink	Russia

Selection: Genetic improvement in pomegranate has, for centuries depended on the selection from seedling variability and their clonal propagation (Pareek and Sharma, 1993). Some of the milestones in the improvement of pomegranate through selection are listed below.

Year	Workdone	References
1932	Seedlings were raised from selected open pollinated fruits of cv. Alandi (deep pink and sour arils, hard seeds) at Ganeshkhind Fruit Experiment Station, Pune in 1932.	Keskar *et al.*, 1993
1936	A promising type, GBG -1 (pinkish and sweet arils, soft seeds) was identified from the seedling progenies of cv. Alandi and released for commercial cultivation in 1936.	Keskar *et al.*, 1993
1970	GBG-1 variety was renamed as Ganesh in 1970.	Keskar *et al.*, 1993
1973	Five promising clones, viz., G-107, G-132, G-133, G-134 and G-137 were identified from vegetatively propagated plants of Ganesh through critical evaluation.	Sawant, 1973
1975	5 better types namely, P -1, P -16, P -23, P -26 and SK -1 were identified from the orchards of Muskat through survey.	Naik,1975
1976-1980	Selection from the seedlings of cv. Muskat and over 47 high yielding individuals superior in fruit quality (soft seeded with high TSS) were identified for further studies. Later a progeny of over 4000 seedling were developed from the selected 47 types and further selections were made. On the basis of yield and fruit quality L26P39, L20P5, L2P19, L26P46, L4P20, L3P38 were selected for better performance (Karale, 1995).	Choudhury and Shirsath,1976; Patil, 1976; Bhapkar, 1976; Karale, 1977; Kolhe, 1980
1982	CO-1, a soft seeded superior pomegranate cultivar was selected from the assemblage of 28 genotypes	Khader *et al.*, 1982
1983	CO-1 was released by Tamil Nadu Agricultural University, Coimbatore	
1984	G-137 with distinctly superior performance for aril colour, aril size and TSS over its parent, Ganesh was released. The performance of five promising selections of Ganesh was further studied.	Keskar *et al.*, 1990
1985	Acc. No. 455, a clone with medium size and easily peelable rind was selected. The seeds are soft with attractive deep purple aril. Later on, this was released as Yercaud -1 due to outstanding performance in Tamil Nadu.	Sayed *et al.*, 1985
1985	Jyoti, a pomegranate with soft seeds and pink aril has been released from University of Agricultural Sciences, Bengaluru. This is a selection for the seedlings of Bassein Seedless and Dholka varieties.	Sulladmath, 1985
1986	P-23 and P -26 were released for commercial cultivation in 1986 after conducting repeated field trails.	Keskar *et al.*, 1993
1996	A new seedless selection RCR -1 from cv. Alandi was reported. It gave 267 fruits per tree in 10 th year with average fruit weight of 220g per fruit and average yield of 58.7 kg fruit/ tree.	Ramu *et al.*, 1996
2013	Selection-4, a promising selection from 62 collections of Bhagwa has been released as Phule Bhagwa Super	Supe *et al.*, 2013

Hybridization: Hybridization is the process of crossing of genetically two dissimilar individuals. The technique of hand emasculation and pollination is followed in pomegranate for development of hybrids. Several crosses were made at Rahuri in 1976, for incorporating blood red colour of Russian types into Ganesh, Out of 122 F_1 hybrids, seven had deep red aril colour but hard seeds and inferior taste than Ganesh (Kale, 1986). Back crossing with Ganesh also did not result in improvement in quality attributes. But, some desirable recombinants in F_2 progeny raised from open pollinated fruits of these hybrids have been identified (Keskar *et al.*, 1989, 1993). A promising selection from the F_2 population (No.61) of Ganesh x Gulesha Red having all desirable quality attributes has been released under the name 'Mridula'.

At IIHR, Bangalore, breeding through hybridization has been going on since 1984. The main objectives are to develop vigorous growing plants having attractive fruits, deep red, bold arils, soft, small seeds and sweet juice. Hybridization among indigenous and exotic pomegranates is the main approach. About 2900 hybrids of single, double, three way and other complex crosses including F2's have been evaluated (Prasannakumar, 1998). The promising progenies have been selected. They not only exhibit heterosis, but combine such attributes like superior fruit quality of Ganesh and Kabul, deep red colour of Gulsha Rose Pink (a Russian introduction) and vigour of Yercaud.

A multiple hybrid has been obtained by involving Ganesh, Kabul, Gul-e-Shah Rose Pink in hybridization. Hybrid No. 15-9-94 has dark red, non-sticky arils and soft seeds with high sweetness and low tannin (Pareek, 1996), which was later released as Ruby. Bhagwa is a selection from the F2 population of the cross Ganesh x Gul-e-Shaha Red.

Hybridization work was initiated at NRCP, Solapur during 2008 with the objective of developing hybrids tolerant to bacterial blight disease and the work is in progress. Few hybrids suitable for processing (*anardana*) purpose and table purpose have been identified.

Mutation : Mutation is defines as the sudden heritable changes in the genotype of an organism. Mutations can be spontaneous (natural) or induced (artificial). The agents used for inducing mutation are known as mutagens. Memedov (1984) treated pollens and seeds with chemical mutagens and observed significant variation in size of the treated pollens and plants raised from the treated seeds.

Levin (1990) treated pomegranate seeds with chemical mutagens and γ- rays and obtained forms with good fruit and juice quality and good keeping quality among mutant seedlings. The best results were obtained with γ-rays at 10-20kR. Promising mutants were also bred by N,N-dimethyl N-nitrasourea treatment. Among the most promising forms produced was the soft seeded seedling 'Sverkhrannil' (meaning Super-Early) with fruit ripening in August. It contains 11-13% sugars, 15% dry matter and 7.2-14.7 mg ascorbic acid per 100g of juice.

Gamma irradiation (3-30kR) of seeds of pomegranate cv. Ganesh and Bhagwa in 2007 has led to the identification of few desirable mutants in pomegranate at NRC on Pomegranate, Solapur.

Important Varieties of Pomegranate Developed in India

Variety	Year of release	Breeding method	Parentage	Institute	Characteristics
Ganesh	1936	Selection	Alandi	MPKV, Rahuri	Soft seeded, yellowish green with red tinge, pinkish arils
CO-1	1983	Selection	-	TNAU, Coimbatore	Soft seeded, high pulp content, sweet taste
G-137	1984	Clonal Selection	Ganesh	MPKV, Rahuri	Soft seeded, yellow with red tinge, deep pink & bold arils, sweet
Jyoti	1985	Selection	Bassein Seedless, Dholka	UAS, Bangalore	Medium-large fruits, Soft seeded, pink arils,
Yercaud-1	1985	Clonal Selection	-	TNAU, Coimbatore	Medium sized fruits, soft seeded, easily peelable rind, deep purple arils
Mridula	1994	Hybridization	Ganesh x Gul-e-shah Red	MPKV, Rahuri	Soft seeded, semi-smooth surface, blood red arils, early maturing
Ruby No.15-5-94	1997	Hybridization	[(Ganesh x Kabul)x Yercaud]—F_1-F_2 x (Ganeshx Gul-e-shah Ros Pink)—F_1-F_2	IIHR, Bangalore e	Soft seeded, red rind and red arils, fruits resemble to Ganesh in size and shape,
Amlidana	1999	Hybridization	Ganesh x Nana	IIHR, Bangalore	Soft seeded, Suitable for processing (anardana),
Phule Arakta	2003	Hybridization	Ganesh x Gul-e-shah Red	MPKV, Rahuri	Soft seeded, dark red rind, dark red arils, early maturing
Bhagwa	2003	Hybridization	Ganesh x Gul-e-shah Red	MPKV, Rahuri	Soft seeded, attractive glossy rind with dark rose pink, bold, red arils
Phule Bhagwa Super	2013	Selection	Bhagawa	MPKV, Rahuri	Early maturing type compared to Bhagwa, superior in quality

Important Varieties of Pomegranate Reported Across Countries

Country	Variety	Characters	Remarks
Iran	760 genotypes & cultivars in Yazd collection		Bahzadi Shahr babaki, 1997
	Malas-e-Saveh Malas-e-Yazdi Rabab-e-Neyriz Sishe Kape-Ferdos Naderi-e-Budrood	Late ripening, medium to large size, thick red rind and red arils	Varasteh *et al.*, 2006
	Alack	Early cv. that ripens in late Aug- early Sep, used for export	Iran Agro Food, 2007
	Alak Shirin (sweet) AlakTorsh (sour)	Red, small sized, hard seeds	-
	Maykhosh	Late, export cv. (till end of Dec)	-
China	87- Qing7	Early bearing, spur type mutant	Liu *et al.*, 1997
	Duanzhihong	Spur type cv. From Xingcheng, ripens in end of August, red skin, pinkish arils, 340g	Liu, 2003
Turkey	Hicaznar	Dark red skin, red arils, sweet/sour	Gozlekci and Kaynak, 1997
	Asinar	505g, large fruit, red arils, sweet-sour, soft seeds	
	Eksilk	Sour(5%TA), red arils	
	Emar	Dark red skin, red arils, sweet with low TA	
	Fellahyemez	Large pink arils, sweet with low TA, soft seeds	
	Katirbasi	517g, large fruit, large red arils, sweet-sour	
Spain	Valenciana	Small, early but not top quality	Costa and Melgarejo, 2000
	Mollar de Elche15	272g, deep red arils with soft seeds, sweet, low acid	Amoros *et al.*, 2000
	Mollar de Orihuela	414g, Red pink arils with soft seeds, sweet low acid	
	PinonTierno de Ojas9	405g,Red pink arils with soft seeds, sweet, low acid	

Contd...

Country	Variety	Characters	Remarks
	Agridulce de Ojos4	524g, Red arils with hard seeds, bitter/ sweet, medium acid	
USA	Golden Globe	Large fruit,Golden green fruit with pink blush, pink to red arils, small soft seeds, sweet	Karp, 2006
	Eversweet	Pink to red fruit with pink arils, soft seeds, sweet even when immature	Dave Wilson nursery, 2005; Karp, 2006
	Wonderful	Deep red arils, medium hard seeds, sweet-sour	Morton, 1987
	Early Wonderful	Deep red arils, medium hard seeds, sweet -sour, 2 weeks earlier than Wonderful	California Rare Fruit Growers, 1997
	Early Foothill	Deep red arils, medium hard seeds, sweet -sour, 2 weeks earlier than Wonderful	La Rue, 1980
	Granada	Deep red arils, medium hard seeds, sweet -sour, 1 month earlier than Wonderful	California Rare Fruit Growers, 1997
Georgia	ApsheronskiiKrasnyi, Burachnyi, Frantsis, KyrmyzKabukh, Lyaliya, Pirosmani, Rubin, Shirvani, Vedzisuri, ImeretisSauketeso	Splitting resistant cultivars	Vesadze and Trapaidze, 2005
Tunisia	Gabsi, Tounsi, Zehri, Chefli, Mezzi, Jebali, Garoussi, Kalaii, Zaghuoani, Andalousi, Bellahi	Gabsi- main cv., sweet; Tounsi - sweet, late ripening, Zehri- Sweet, ripens end of August-Sep, Chefli - Sweet, bold arils; Mezzi, Jebali, Baroussi – Sweet-sour, green skin; Kalaii- Sweet, bold arils; Zaghuoani, Andalousi-sweet	Mars and Marrakchi, 1999
Egypt	Arabi, Manfaloty, Nab ElGamal&Wardy	Manfaloty- sensitive to salt stress; Nab ElGamel - Saline tolerant	Abu-Taleb et al.,1998; Saeed,2005

Contd...

Country	Variety	Characters	Remarks
Iraq	Ahmar (red), Aswad (black), and Halwa	Ahmar-red, Aswad -black	Morton, 1987
Saudi Arabia	Mangulati		Morton, 1987
Vietnam	Vietnamese	Evergreen cv. With orange flowers, bright red skincolour and small, juicy arils	Jene's Tropical Fruit, 2006
Morocco	Gjeigi, Dwarf Evergreen, Grenade Jaune, Gordo de Javita, Djeibali, OnukHmam	17 clones and cultivars were reported	Oukabli et al.,2004
Sicily, Italy	Dente di Cavallo, Neirana, Profeta, Racalmuto, Ragana, Selinunte	6 selections in Sicily	Barone *et al.* 2001

Achievements of NRCP, Solapur

In collaboration with IIHR, Bengaluru, hybridization and hybrid evaluation work was initiated in 2007-08. Few hybrids were identified for table purpose besides few hybrids for anardana purpose. Rigorous screening of pomegranate hybrids developed at IIHR, Bengaluru has paved the way for identification of some pomegranate hybrids which are tolerant to bacterial blight disease caused by *Xanthomonas axonopodis* pv. *punicae*. These bacterial blight tolerant hybrids could be useful as parents in further hybridization programme.

- Hybrids identified for table purpose: NRCP H-6, NRCP H-1, NRCPH-14
- Hybrids identified for *anardana* purpose: NRCP H-4, NRCP H-12
- Hybrids identified for bacterial blight tolerance: NRCP H-1, NRCP H-2, Bhagwa x [(Ganesh x Nana) x *Daru*], Nayana x Ruby

Conclusion & Future Thrust Areas

It is high time to identify resistant source for bacterial blight disease and development of bacterial blight resistant varieties/ hybrids besides development of wilt resistant varieties has become the need of the hour and deserves top most priority in the crop improvement arena of pomegranate. Anther culture technique is useful for the development of haploid plants which is homozygous in nature. Doubled haploidy (DH) technique has to be exploited in pomegranate through colchicine based chromosome doubling of haploid lines. Identification of suitable markers and work on Marker Assisted Selection (MAS) would contribute for the improvement of pomegranate.

3

The Quest to Develop Pomegranates Cultivars for Alimentation, Sustenance and Entrepreneurship: Intricacies and Accomplishments

B.N.S. Murthy

ICAR-Indian Institute of Horticultural Research, Bangaluru-560 089, Karnataka
E-mail: bnsmurthy@yahoo.com

Introduction

Pomegranate (*Punica granatum* L.) belongs to family Lythraceae with 2n=16 or 18 chromosomes. *Punica* is a small genus of fruit-bearing deciduous shrub or small trees. The genus *Punica* has only two species *viz.*, *granatum* and *protopunica*. While, *Punica granatum* has several commercially cultivated varieties, *Punica protopunica*, the Socotra pomegranate is endemic on the island of Socotra (part of Yemen) it differs in having pink (not red) flowers and smaller, less sweet fruits with mesophytic habit against the xerophytic habit of *P. granatum*.

Pomegranate occupies the eighteenth place among the main world-fruit cultures. This ancient fruit crop is native to hot dry regions of Afghanistan and Baluchistan, but with time it has diffused and got adapted to a wide range of climatic conditions. The unique plasticity of this fruit crop is evident from the threshold limits it exhibits for high (44°C) and low (-12°C) temperatures and several frost-hardy varieties that can stand even as low as -10°C. It travelled to Central and South India from Iran around 1st century AD and was reported growing in Indonesia in 1461. The most important growing regions in the world are Egypt, Spain, Turkey, Morocco, Tunisia, Georgia, China, Afghanistan, Pakistan, Bangladesh, Iran, Iraq, India, Saudi Arabia, Turkmenistan and Tajikistan. It has also been found place in Israel on the coastal plains and Jordan valley. In the Western world, pomegranate cultivation was first recorded in 1621 in West Indies and warmer areas of South and Central America, from there it was introduced into California in 1769 by Spanish people. It produces fruits in tropical, sub-tropical, temperate regions and in hilly tracts up to 1800 m altitude. The plant habit, fruiting and flowering physiology are, however, altered with the habitat. In the tropical climate of South India with a mild winter, the growth and flowering is continuous throughout the year, while in the sub-tropical climate of North India, the trees remain dormant during cold winter and the main flowering occurs in the following spring, whereas under the temperate climate of Himachal Pradesh flowering will be in summer. Thus the plant habit, fruiting and flowering physiology are altered with the habitat.

Objectives of Breeding

Although, until last decade the well-known objectives in any pomegranate genetic improvement programs included improving the traits like organoleptic qualities, juice content, seed mellowness, appearance of fruits in addition to fruit yield, the focus is now turned to biotic and abiotic resistance introgression. While, among the biotic challenges, bacterial nodal blight and wilt are the major focus, physiological disorders like fruits that do not exhibit cracking and aril browning are receiving importance under abiotic resistance breeding. Further, since the preferences of the fruit traits were not same in all countries, development of new varieties were mostly targeted for indigenous markets and to a limited extent to export market. However, the scenario is changing now a days as pomegranate has become one of the economically important fruit crop and is being exported to distant markets. In such a case characters related to international markets like fruit size, skin colour, aril colour, sugar acid blend and beneficial nutrients in the juice, besides ripening time and shelf life of the fruits are being considered by breeders. The other desirable traits relating to flowering (synchronised and increase in ratio of the production of hermaphrodite to male flowers) and plant canopy attributes (less thorny and upright growth habit), high temperature, frost, and moisture stress tolerance are also receiving importance worldwide in pomegranate improvement programmes.

The fruit standards to be considered while making selection include fruit size of over 400g, red glossy and thin rind (<0.25 cm), dark red or dark pink juicy arils (over 70%), bold arils each weighing 400 mg and above, high TSS (> 16° B), less acidic (< 0.4 %), very soft seeds (< 2 kg/cm^2), good shelf life (> 10 days) and lower levels of tannins (<150 mg/100ml juice) *etc.*

Varietal Status

In almost all countries where pomegranate is commercially grown, despite the availability of large number of local varieties only few are commercially utilized. The names of the cultivars originate frequently either from the place of cultivation or from the colour of the fruit. Varieties are often classified as sweet, sweet-sour and sour, early, mid-season and late, juicy and table fruit, soft seeded and hard-seeded *etc.*

Several cultivars grown today are the result of human selection from naturally occurring variation. There are more than four hundred cultivars of pomegranate differing in the habit of the tree, leaf type, fruit form, size, colour, aril characters, keeping quality *etc.*, In India until last decade, Ganesh variety was by far the most popular one. This is a seedling selection from a hard seeded 'Alandi'. It produces large size (400-450g), fruits with sweet (16-17 °B) arils containing soft seeds. But the arils are pink or light pink in colour. Hence, in the last decade varieties of pomegranate *viz.*, Ruby, Arakta and Mridula with red aril colour were developed using Ganesh variety as base. Both the varieties derive genes for red aril colour from Russian temperate varieties. Ruby is a multiple hybrid resembling more of Ganesh, but for the red aril and fruit colour, while Arakta is a F_2 selection for dark red arils. In recent years, the ruling commercial variety, covering an area of over 90 per cent with good eating qualities and shelf life-because of thick skin and is being called by various names as Bhagwa, Astagndha, Mastani or Keshar, Sinduri *etc*. This variety is a selection from segregating hybrid progeny of unknown parents and has gained importance among the growers quickly though it is slow to mature (about 160 to 180 days). In addition to this, there are several other seedling selections which are grown on a limited scale across the country like Yercaud, CO-1 (Tamil Nadu), Dholka (Gujarat), Jalore Seedless, Jodhpur Red (Rajasthan), Muscat (Maharashtra) and Panji (Goa). In Himachal Pradesh a sour pomegranate type-Daru, comes abundantly in wild. This is used mostly in the preparation of anardana, an acidulant product used in the culinary preparation. Kabuli, Kabul Yellow, Paper shell, Spanish Ruby and few others, although are known for some time, they have not achieved prominence. Several temperate introductions like Gulsha Rose Pink, Kali Shirin, Kazaki Anar, Lupania, Shirin Anar, Sunni Bedana *etc.*, do not have commercial significance in tropics.

Approaches

Pomegranate is considered as a monoecious species and develops three kinds of flowers; male flowers-with short styles and atrophied ovaries containing few eggs, which are generally, bell-shaped; hermaphrodite flowers -with normal ovary developing to fruit, which are, in general, peanut shaped and the third category is intermediates. The hermaphrodite flowers exhibits protandry-dichogamy, where female and male structures do not mature simultaneously. Inadequate pollination is one of the most important factors that limit commercial production resulting in low fruit set complained from farmers. From several studies it is implied that pomegranate is an often-cross pollinated fruit crop and it is highly heterozygous, does not produce true to mother progeny and segregation occurs in F_1 generation itself. Adequate variation for several plant and fruit traits is generated in nature as well as upon crossing. Although the present day commercial orchards of pomegranate are either from layers or cuttings, in olden days people resorted to seed propagation as seeds germinate readily and often produce vigorous plants Hence, cultivation of pomegranate over centuries and seed propagation has resulted in generating great variation and subsequently in development of several varieties, highly local in their adaptation.

Improving the appearance and organoleptic qualities and shelf-life of fruits, selecting cultivars or rootstocks that tolerate biotic and abiotic stresses, reducing the size of the plant in order to increase orchard plant density, selecting self-fertile genotypes to maintain a more consistent yield over time and genotypes with higher nutritional value of the fruit are paramount desirable traits of pomegranate and this may be brought about by any one or combination of the following approaches.

- Clonal / Seedling Section
- Introduction of Varieties
- Hybridization- Sib-mating, Intra-specific, Inter-specific (limited scope), Single cross,
- Multiple cross, Selection in F2, Sib mating, Backcrossing
- Mass Selection
- Mutation / Tillage
- Polyploidy Breeding
- Marker Assisted Selection
- Transgenics

Breeding objectives are generally achieved by evaluation of traits amongst genetic resources and eventually improve the tree parameters by collecting desirable characters in one cultivar. Hence, the prerequisite is to have wide genetic base. Some of the statistical methods including principal components or cluster analysis can be used for screening accessions. This would help in elucidating genetic relationship among pomegranate genotypes that can be used for selection of parents in breeding programs.

Most of the pomegranate varieties in cultivation were often developed by seedling or clonal selection by mainly exploiting the inter-varietal variability as plants are highly heterozygous and produce heterogeneous population with pronounced differences in productivity and vegetative characteristics and give adequate scope for desirable selections. There is large variability among cultivars for fruit size and shape, rind and seed colour, juiciness, sugar content and acidity, taste *etc.*, and some of them may be strongly influenced by the environment. In this scenario, one should be cautious while using morphological or physiological characters as selection criteria. Obfuscating to this situation is absence of definite correlation between either fruit size or extent of skin red coloration and internal fruit composition. Further, when wild / non-cultivated types are used in the hybridization work (that is often the case) in order to eliminate undesirable traits in the progenies one has to go for repeated backcrosses by making selection in each generation which is both time and resource consuming.

One of the tools which are making it easier for scientists to select plant traits and develop new varieties is the use of molecular markers. In perennial crops like pomegranate it is very useful tool in breeding, where the breeder has to wait for several years to see the expression of the desired trait. This approach therefore needs much attention which will help in early selection of desirable lines.

The space-induced mutation technique, so called space breeding is a crop breeding technique which uses the good variations of plants (seeds) induced in the space environment that can be reached by the recoverable spacecraft (such as recoverable satellites and space shuttles) and high altitude balloon to choose new germplasms and new materials on the ground, then to develop new crop varieties. Aerospace provides a special environment with strong cosmic radiation, microgravity, weak geomagnetic field and super-vacuum, *etc.*, which might affect plant growth and development as well as induce genetic changes of crop seeds. Space breeding is a novel effective approach to crop mutational improvement and it is perceived that space-induced mutation breeding is an effective way to both breed new varieties and enhance genetic diversity. It is believed that crop space-induced mutation breading can be a novel effective way to both breed new varieties and create distinctive genetic resources because of its wide mutation spectrum, high frequency of useful genetic variation as well as short breeding period. This approach is being attempted at IIHR, in collaboration with ISRO.

Mutation breeding has some limitations, as it is only a 'hit and miss' method and needs rising of large population to get a desirable type. Nevertheless, the 'Targeting Induced Local Lesion In Genome' (TILLING) is a powerful technique for identification of mutations and elucidation of gene function for traits of interest. TILLING, a non-transgenic reverse genetics approach is envisaged to develop resistance against bacterial blight of pomegranate. This technique has been successfully applied to large variety of plants including pea, tomato, soybean, maize *etc.* for multiple traits including disease resistance. It has evolved as an alternative, non-GM strategy to introduce novel alleles for target traits. TILLING of genes putatively involved in primary infection process by pathogen for *e.g. Xanthomonas axonopodis* pv. *punicae* may help in recovering novel alleles which may prevent host recognition and establishment. Development of transgenic varieties is yet not endeavoured in pomegranate. Attempts have been initiated to transfer the resistant gene analog *Xa21* (confers resistance to *X. oryzae* in rice) from rice, however it is unlikely that *Xa21* a protein kinase family gene confers resistance *X. axonopodis* pv. *punicae.*

The tropical and temperate pomegranates are expected to contain a distinct gene constellation as they have followed a contrasting evolutionary pathway in separate geographic regions with different environmental conditions. Since the gene transfer across varieties of diverse origins is straight forward in pomegranate, it may be worth using them in breeding for several other plant and fruit attributes. Further, despite inherent narrow genetic base that exists in pomegranate, there is great scope for selecting the desirable genotype this is evident from the extensive perused study carried out at IIHR, Bangaluru. Among the progeny of 35 crosses studied over five generations, wide segregation was noticed with extreme forms having fruit as small as 39g and as big as 658g and TSS as low as 8.6°B and as high as 18.8°B. Likewise, recombinants appeared with a gradation of fruit skin and aril colour, while seed mellowness ranged from very soft to very hard. This breeding program did result in developing of a new multiple pomegranate hybrid Ruby.

Large flowers, abundant seed per cross, their high germination percentage, relatively short juvenile phase of seedlings make pomegranate breeding slightly easier as compared to breeding of several other fruits crops. The commercial practice of pomegranate propagation through air layers or cuttings facilitates in fixing of beneficial characters identified in a given genotype whether it is governed by a major gene or group of modifiers or pseudo-over dominance. The disadvantage of the breeding program programme is that it is time consuming as several undesirable traits have to be eliminated and the screening of the hybrids needs to be carried out stringently both in the seedling stage in the glass house and in adult stage in field in hot spots in case of biotic resistance breeding programs.

Challenges

Genetic improvement of pomegranate faces range of obstacles. These include the narrow genetic base (absence of related species), long juvenile period, frequent intra-varietal incompatibility, high level of heterozygosity, sterility and the presence of specific traits only in wild relatives. These physiognomies make breeding techniques difficult, expensive and time consuming. This explains why there are mostly clonal selections, almost exclusively using variability from spontaneous mutations or selecting plants derived from natural hybridization. Recent developments, such as induced mutations by ionising irradiation, have given few promising results both for cultivars and rootstocks in other crops. Molecular and biotechnological approaches such as soma-clonal variation or gene transformation, protoplast technology, for example, provides the potential for making significant changes to varieties, but limited attempts have been made in pomegranate using such tools.

Genetics of Important Traits

Success in breeding for economically important traits depends upon the information available on the gene action governing the inheritance of the traits concerned and planning and execution of breeding programs accordingly. Thus knowledge on inheritance of traits is a pre- requisite in applied pomegranate breeding. Deliberate efforts to develop pomegranate varieties with specific objectives have not received adequate attention. A non-overlapping unambiguous grouping is necessary in the inheritance studies. But most of the fruit crops including pomegranate they exhibit long reproductive cycle, high level of heterozygosity and difficulty in raising massive population therefore studies on genetics of fruit crops are limited. Moreover, several characters are highly sensitive to environmental fluctuations making inheritance studies much more difficult. There are classical examples depicting environment effects masking and confounding the subtle differences in expression of traits due to genetic factors. The circumscribed studies that have been carried out to understand the genetic bases of various traits in pomegranate hint at the following gene action.

- **Flower type:** Monogenic and double flower is dominant to normal, while sex expression is yet to be divulged.
- **Fruit skin colour:** Polygenic and highly sensitive to environment, especially sunlight. Fruit size: Large fruit size is dominant to small fruit size.
- **Foliage, flower and fruit colour:** Monogenic and red/pink is dominant to yellow.
- **Aril colour:** Red /pink aril colour is dominant to light pink / white and temperature sensitive.
- **Fruit acidity:** Monogenic with modifying genes; high acidity is dominant over low acidity and a linked character with seed hardiness.
- **Seed mellowness:** Polygenic; hard seediness is dominant over soft seeded types. Rosette plant type: controlled by a single recessive gene.
- **Bacterial blight resistance:** Suspected recessive trait.

Breeding for Biotic Resistance

In recent days, pomegranate has emerged as a commercially important fruit crop in the Indian arid tracts. The production of this fruit has attracted several growers for its low cultivation cost, drought tolerance and export potential. The fruit has wide consumer preference for its attractive, juicy, sweet-acidic and refreshing arils and there is a growing demand for good quality fruits both for fresh use and processing into juice, syrup and wine. The area under this crop is fast increasing. At the same time, bacterial nodal blight (BNB), a devastating disease, has attained serious proportion that some of the growers are uprooting the trees due to lack of resistant variety and

/or effective control measures. All the important varieties cultivated in tropics are prone to BNB and there is an urgent need to develop resistant varieties to defend the pomegranate industry from collapsing. The disease caused by *Xanthomonas axonopodis* pv. *punicae* is serious and affects flowers, fruits, leaves and twigs. Plant disease management should be one of the main objectives of any crop improvement program as the most effective form of disease control is host plant resistance and should receive priority. Growing of pomegranate plants with built in tolerance is economical, environmental friendly and ecologically sound. The main thrust in pomegranate breeding until recently was to develop varieties with dark red arils and soft seeds. For good fruit quality the indigenous cv. Ganesh, while for aril colour, the exotic Russian types served as useful sources of genes now because of the enormity of BNB problem the focus of the breeders is shifting towards biotic resistance breeding utilizing resistant sources.

The breeding program carried out at IIHR, Bangaluru, has revealed high level of susceptibility in the progenies of several crosses, back crosses, multiple crosses involving tropical types and Russian temperate types which suggested the lack of resistant genes, and a need for isolation of resistant source from other groups. In another study conducted for assessment of genetic diversity in the wild pomegranates obtained from different sub-temperate regions of the country (locales of Western Himalayan region, Himachal Pradesh and Uttarakhand) using DAMD and RAPD profiles, it was found that these wild genotypes come up naturally in abundance, and these hardy deciduous seedling trees growing since long time possess better climatic adaptability and tolerance to pests and diseases (owing to natural selection over several years) were distinct. Further, these could be useful for identifying new germplasm sources that, when crossed with existing commercial varieties would result in imparting tolerance to different kinds of stresses. Another variety 'Nana', a miniature pomegranate (like bonsai) having ornamental value, also exhibits tolerance to bacterial blight. Since, there is great scope for gene exchange between edible and ornamental pomegranates they are valuable donors for bacterial blight tolerance and show a ray of hope in developing tolerant types. But, in order to transfer the targeted resistant genes and to eliminate the undesirable traits of the donors (small fruit size, high acidity and hard seediness *etc.*) selections over few generations would be necessary. Further, two Sri Lankan varieties Kalpitya and Nayana with hard seeds and light pink arils have also been identified to possess tolerance for bacterial blight. When they are used in the crossing work as sources of tolerance, the number of traits to be eliminated is far less than from the non-edible types i. e. *Daru* and Nana.

The other approaches envisaged for development of BNB resistant lines include the path of androgenesis, the process of induction and regeneration of haploids and double haploids originating from male gametic cells. This is relevant in the light of the report that resistance to BNB is controlled by recessive gene and selection process in the diploid phase would be challenging. Pollen derived haploid plantlets from unrelated crosses be screened for *BNB* and rated as resistant, intermediate or susceptible. The resistant or tolerant haploid plants should then be colchicine-treated to produce doubled haploid (DH) lines. Due to its high degree of effectiveness and applicability in numerous plant species, this approach has outstanding potential for plant breeding and commercial exploitation in pomegranate. The next approach is pollination with irradiated pollen; which is another possibility for inducing the formation of maternal haploids using intra-specific pollination. Embryo development is stimulated by pollen germination on the stigma and growth of the pollen tube within the style, although irradiated pollen is unable to fertilize the egg cell. It has been used and successfully demonstrated in several species including fruit crops.

The fungal wilt caused by *Ceratocystis fimbriata* in association with the shot hole borer (*Xyleborus perforans* Wollastan.) and nematodes is also attaining serious proportion in several pomegranate growing areas. Researchers are yet to establish the genetic differences for the resistance to wilt; which is a pre-requisite to identify the source of resistance. Once that is established, there will be scope for development of rootstocks that can be used to counter the attack of the causal agents as no satisfactory chemical control measures are available at present. Screening for *Ceratocystis fimbriata* has to be done in those genotypes that exhibit high level vigour and stress tolerance by raising the seedlings in the sick soil or by artificial screening by dipping seedlings in spore suspension of pathogen after wounding the roots and then planting them in pots containing sterile soil. Work in this direction is already taken up at NRC on pomegranate, Solapur and IIHR, Bangaluru. The *Daru* seedling population may be a good beginning for this purpose as they are hardy in nature and graft compatible with commercial pomegranate cultivars.

Breeding for Abiotic Resistance

Aril browning and fruit cracking are the major abiotic torments that need immediate attention. In a preliminary study at IIHR, Bangaluru, in order to understand the genotypic variation for aril browning and its association with other fruit traits, a total of 168 multiple pomegranate hybrids were scored for fruit colour, presence of beak, aril colour, seed softness, TSS and for aril browning. Aril browning was scored as A (No browning), B (High), C (Medium), D (Less), and E (Very less). Because of diverse genetic base of parents and heterozygous nature of the crop a wide array of recombinants were produced. Spearman's correlation analysis revealed that aril browning is inversely related with skin and aril colours. A statistical model constructed to study the reasons for the observed variation in aril browning showed with the increase in intensity of skin colour and aril colour there will be reduction in severity of aril browning while with raise in TSS aril browning incidence will be more, an association not favourable in selection of desirable genotypes. Hence, striking a balance in aril colour and TSS level while locating recombinants free from aril browning is necessary.

Attainments

Cultivation of a good variety is of prime importance as the success of a crop hinges on the variety used. Although over one hundred pomegranate varieties have been developed for commercial cultivation over ages in different parts of the world only a few could survive the competition and the scenario is changing often. CO-1, a variety developed in Coimbatore in 1976, is a good yielder of attractive fruits with soft seeds. Some of the popular varieties of pomegranate in India are Alandi or Vadki, Dholka, Kandhari, Kabul, Muskati Red, Spanish Ruby, Bassein Seedless, Jyothi, Ganesh, G-137, P-23, P-26, and Yercaud-1. Most of them are of seedling origin from the varieties introduced from neighbouring countries. Ruby and Mrudula (Arakta) are hybrid pomegranates developed by IIHR and MPKV respectively. Both the hybrids possess deep red arils and they derive the genes for this pigment from temperate CIS cultivars. They have soft seeds with brownish red to red skin colour. Amlidana is a F_1 hybrid variety developed by IIHR and is good for anardana making. In the recent years Mridula, Bhagwa, Super Bhagwa are popular amongst farming community. Given below is the list of popular pomegranate cultivars across the globe developed using different breeding techniques.

- 87-Qing 7 — A natural mutant of cv. Qingpitian
- Aleko — Developed through hybridization
- Azerbaijan — Developed through hybridization Purpurovyi [Purple] X Krmyzy kabukh
- Baiyushizi — Bud sport of cv. Sanbai
- Bhagwa — Selection from segregating hybrid progeny of unknown parents.
- Desertnyi (Dessert) — Developed through hybridization (Wonderful X Soviet cvs.) X Wonderful
- G-137 — Clonal selection from cv. Ganesh
- Hongyushizi — Mutant of cv. Yushizi
- Mengliaihong — Seedling selection
- Mridula — Selection in F2 in a cross Ganesh X Gul-e-Shah Red
- Nasimi and Vurgun — Developed through hybridization
- RCR1 — Seedling selection from cv. Alandi
- Ruby — Selection in F2 in a cross Ganesh X Gul-e-Shah Red
- Shani-Yonay — Seedling selection
- Taishan Dahongsh-iliou — Chance seedling
- Zaoxuan018 — Seedling selection from cv. Dahongpao
- Zaoxuan 027 — Seedling selection from cv. Daqinpi.

4

Molecular Markers in Genetic Improvement of Pomegranate

Kanupriya, K.V. Ravishankar and B.N.S. Murthy

ICAR-Indian Institute of Horticultural Research, Bengaluru-560 089, Karnataka

Genetic improvement of fruit crops is challenging since trees are woody perennials with a long breeding cycle and heterozygous nature of the genome. Fruit trees are of great importance in developing countries since most of them produce highly nutritious fruits which are major sources of vitamins for human beings. Value of these crops in providing quality nutrition and for uplifting economic condition of the farmers is being increasingly recognized. One such fruit crop is pomegranate. Pomegranate (*Punica granatum* L.; 2n=2x= 16/18) is a diploid, perennial, woody plant and belongs to the monogeneric family Lytheraceae. The genus *punica* possesses two species *viz.*, *Punica granatum* L. and *Punica protopunica* Balf. The genome size of pomegranate has been ascertained to be 704 Mbp, about six times the size of *Arabidopsis thaliana* (Bennette and Leitch, 2010). This fruit is believed to have originated in Iran (Levin, 1994) from where it diversified to other regions like Mediterranean countries, India, China, Afghanistan, through ancient trade routes. Pomegranate, which is a rich source of polyphenols, tannins and anthocyanins, is included among a novel category of plant sources called 'superfruits'. Commercial potential of this crop can be greatly enhanced, if improved cultivars with resistance to biotic and abiotic stresses can be developed with features that meet consumer preference. The role of molecular markers in breeding of fruit crops is being increasingly realised, since conventional fruit breeding is largely based on phenotypic selection with the help of morphological markers; however, this takes years to be usable since, fruit trees have a long generation time and are mostly cross pollinated. Also these markers are subject to environmental factors. Traits of economic importance like fruit quality, yield, precocity and disease resistance are polygenic and complex in nature. Therefore, advanced genetic tools like molecular markers (which can be used on any tissue at any time of plant growth, overcoming limitations of traditional methods) are being identified and used in breeding/ varietal identification of fruit trees.

The molecular markers have been used for the evaluation of genetic diversity in pomegranates. Dominant markers have been used by several investigators to conduct evaluations of genetic diversity like randomly-amplified polymorphic DNA (RAPD) by Dorgac *et al.* (2008), Sarkhosh *et al.* (2006), Talebi *et al.* (2003), Zarei *et al.* (2009), inter simple sequence repeats (ISSR) by Talebi *et al.* (2005), amplified fragment length polymorphism (AFLP) by Jbir *et al.* (2008), La Malfa *et al.* (2010), Yuan *et al.* (2007), Rahimi *et al.* (2003), sequence related amplified polymorphism (SRAP) by Soleimani *et al.* (2012) and directed amplification of minisatellite DNA (DAMD) by Narzary *et al.* (2009).

In most of these studies, the grouping by the marker profiles does not agree with the grouping based on morphological traits and geographical origin. Some studies have reported a high degree of similarity among the pomegranate genotypes (Talebi Baddaf *et al.*, 2003) while others have reported high polymorphism among them (Sarkhosh *et al.*, 2006; Jbir *et al.*, 2007; Narzary *et al.*, 2009).

With the development of microsatellite markers in 2007 by Koohi *et al.* were the first to report the isolation of 15 microsatellites in pomegranate, the scenario started to change. Microsatellite markers have been used in many fields like genetic resource conservation, establishment of core germplasm, population genetics, molecular breeding and fingerprinting of varieties. The wide range of application of these markers is due to their codominant, multiallelic and highly reproducible nature; high resolution; amenability to high-throughput and polymerase chain reaction (PCR). These markers have been discovered traditionally by constructing genomic libraries enriched for a few, targeted SSR motifs and sequencing of clones containing SSRs. However, such library based approaches are expensive, labour and time consuming and can isolate only the targeted, enriched SSR motifs (Zalapa *et al.*, 2012). Available genomic resources like ESTs are also used for isolating SSRs *in silico*. However, EST derived markers are mostly monomorphic; thus, they are not generally useful for developing linkage maps and for DNA fingerprinting.

A large number of SSR markers and mapping populations is required for developing a high-density linkage map. Availability of such maps along with phenotypic data helps in identifying QTL regions associated with traits of interest. Once such associations are identified, application of MAS to fruit crops is accelerated.

Few studies have been conducted on marker trait associations in pomegranate using microsatellite markers. Basaki *et al.* (2013) evaluated the relationship between microsatellite markers and important traits like seed softness and aril size by using multiple stepwise linear regression analysis.

In pomegranate, SSR and SNP markers have been developed in recent years using the next generation sequencing technologies (NGS). Using Illumina platform candidate genes for hydrolyzable tannin, anthocyanin, flavonoid, terpenoid and fatty acid biosynthesis and/or regulation along with 115 SSR markers were identified in the cDNA developed from pomegranate fruit peel (Ono *et. al.* 2011). In another study by Ophir *et al.*, 2014, two contrasting genotypes of pomegranate 'Nana' and 'Black' were used for construction of transcriptome library. Using 454-GS-FLX pyrosequencing technology a resource of 7,155 simple-sequence repeats (SSRs) and 6,500 single-nucleotide polymorphisms (SNPs) was developed. A subset of SNPs was used to assess the genetic diversity in 105 pomegranate accessions from different geographical regions of the world viz., including India, China, Central Asia, USA, Spain, Turkey and the Mediterranean region. The genetic classification was found to divide the germplasm collection into two statistically significantly distinct genetic groups: G1 and G2. The G1 group was composed of *P. granatum* var. Nana seedlings and its descendant accessions, in addition to accessions of Indian, Chinese and Iranian origin. It included the Indian cultivar Bhagwa and all of the evergreen accessions. The G2 group included accessions from the Mediterranean region, Central Asia and California. The pomegranate accessions were separated into two general geographical regions. One branch spread from the suggested origin of the pomegranate species toward the Far East, while the other spread toward the West. This study provides a broader information about the germplasm diversity since it is based on large set of genetic markers and accessions.

Attempts have been made at Indian Institute of Horticultural Research, Bengaluru to broaden the genomic resources in pomegranate for supporting breeding strategies and for genetic mapping. Making use of NGS technologies, large numbers of microsatellite markers have been developed (Ravishankar *et al.*, communicated). Using the NGS data, gene annotations and miRNA mining has been done. Five miRNAs were identified in pomegranate using a bioinformatics approach on 7,361 contigs from partial genome sequence data (19 Mbp) generated using Roche 454 GS FLX Titanium pyrosequencing technology. Thirty-five potential mRNA targets were identified by homology searches of all five identified pomegranate miRNAs against the *Arabidopsis thaliana* mRNA dataset using psRNAtarget software (http://plantgrn. noble.org/psRNATarget/). The miRNA targets identified were then subjected to Gene Ontology analysis. We found that most mRNA targets were constitutively expressed genes involved in different molecular functions, biological processes, and cellular components. Experimental validation of these computationally-identified miRNAs has the potential to elucidate miRNA-based gene regulation and evolution in a non-model crop such as pomegranate (Kanupriya *et al.*, 2013).

It is expected that in coming years the molecular markers developed for this crop will be used to identify interrelations among pomegranate accessions, development of linkage maps, marker-trait associations and a better understanding of genetic variation present in pomegranates.

5

Guidelines to Conduct Test for Distinctiveness, Uniformity and Stability (DUS) in Pomegranate (*Punica granatum* L.)

Ram Chandra, Swati Suryawanshi, Rigveda Deshmukh, Swapnil Sawant, Umesh Birajdar and R.K. Pal

ICAR-National Research Centre on Pomegranate, Solapur - 413 255, Maharashtra
E-mail: rkrishnapal@gmail.com

Introduction

India as a member of the world trade organization (WTO) is obliged to comply with the Agreement on Trade Related Aspects of Intellectual Property Rights (TRIPS Agreement). This requires that member countries provide Intellectual Property Rights (IPRs) for protection, patenting and commercialization of technology. In order to protect plant variety and farmers rights of pomegranate, guidelines along with descriptors are essentially required. In the recent past, Protection of Plant Variety and Farmers Right Authority (PPV & FRA) has developed guideline for protection of pomegranate varieties including extant one. This document describes complete guidelines including material required, procedure for conduct of DUS test , methods and observation, grouping of varieties, characteristics and symbols and DUS testing centres in India. With the help of guidelines breeders and farmers involved in pomegranate can register their variety (ies) and get protection by PPV & FRA.

Guidelines for the Conduct of DUS Test

A. Plant Material Required

1. The PPV & FRA shall decide on the quantity and quality of the plant material required for testing the variety and when and where it is to be delivered for registration under the PPV & FR Act, 2001. Applicants submitting such plant material from a country other than India shall make sure that all customs and quarantine requirements stipulated under relevant national legislations and regulations are complied with.

2. One year old plant material for testing is to be supplied in the form of 10 propagules such as air-layered plants/rooted stem cuttings (multiplied from the same tree)/ tissue culture raised plants, *etc.* for each location.

3. The plant material supplied should be healthy, vigorous and not affected by any important insect pests or diseases.

4. The plant material should not have undergone any treatment, which would affect the expression of the characteristics of the variety, unless the competent authority allows or directs for such treatment. If it has been treated, full details of the treatment must be given.

B. Conduct of Tests

1. The minimum duration of the DUS tests shall normally be at least two similar fruiting seasons in different years. The tests shall be conducted at least at two locations.

2. The tests should be carried out under conditions ensuring satisfactory growth for the expression of the DUS characteristics of the variety and for the conduct of the examination.

3. The test design for the testing should be such that plants or parts of plants may be removed for measurement or counting without prejudice for the observations which must be made up to the end of the growing cycle.

C. Methods and Observations

1. The characters described in the table of characteristics [section F] shall be used for the testing of varieties for DUS.

2. For the assessment of distinctiveness and stability, observation shall be made on at least five plants (multiplied from the same tree) or parts taken from each of 5 plants. In the case of parts of plants, at least two parts should be taken from each of the five plants.

3. Observations on tree or one year old shoot of the tree should be taken at the end of crop season.

4. Observations on mature leaf should be taken from the one third portion of the current season's shoot from the apex emerged on middle branch.

5. Observations on the flower should be recorded on the hermaphrodite flowers when completely open.

6. Observations on the fruit should be recorded on five fruits selected randomly from all directions of the tree canopy, on fully mature/completely ripened fruits ready for consumption.

7. Observations on the peel (rind) should be recorded from the equatorial zone of the fruit.

8. Observations on the seed and arils should be recorded from the fresh seeds and arils.

9. Total soluble solids (TSS) should be recorded by hand refractometer (0-32⁰ Brix).

10. Acidity in fruit juice should be determined by titration against N/10 NaOH using Phenolphthalein indicator.

11. Fruit juiciness should be recorded on fresh weight basis of total fruit weight and total juice extracted from the fruit.

D. Grouping of Varieties

1. The candidate varieties for DUS testing shall be divided into groups to facilitate the assessment of distinctiveness. Characteristics, which are known from experience not to vary, or to vary only slightly within a variety and which in their various states are fairly evenly distributed across all varieties in the collection are suitable for grouping purpose.

2. Grouping of characteristics are those in which the documented states of expression, even when produced at different locations, can be used, either individually or in combination with other such characteristics to (a) select varieties of common knowledge that can be excluded from the growing trial used for examination of distinctiveness; and (b) organize the growing trial so that similar varieties are grouped together.

The following characteristics are to be used for grouping pomegranate varieties:

a. Flower   : colour of calyx (characteristic 14)

b. Fruit shape   : ratio of longitudinal and lateral axes (characteristic 21)

c. Ripe fruit   : colour (characteristic 22)

d. Seed   : hardiness (characteristic 30)

e. Aril   : colour (characteristic 27)

f. Fruit maturity : days after anthesis (characteristic 33)

E. Characteristics and Symbols

1. To assess distinctiveness, uniformity and stability, the characteristics and their states as given in the table of characteristics [Section F] shall be used.

2. Notes (1 to 9) shall be given for each state of expression for different characteristics for the purpose of electronic data processing.

3. Legend (*) Characteristics that shall be observed during every growing season in all varieties and shall always be included in the description of the variety, except when the state of expression of any of these characters is rendered impossible by a preceding phenological characteristic or by the environmental conditions of the testing region. Under such exceptional situation, adequate explanation for such characters shall be provided. (+) See explanation on the table of characteristics in Section H. It is to be noted that for certain characteristics, the plant parts on which observations to be taken are given in the explanation or figure(s) for clarity and not the colour variation.

4. The optimum stage of plant growth for assessment of each characteristic is given in the 6th column of the table of characteristics [Section F].

5. Types of assessment of characteristics indicated in column seven of table of characteristics [Section F] is as follows :

    MG : Single measurement of a group of plants or their parts.
    MS : Measurement of number of individual plants or their parts.
    VG : Visual recording of single observation of a group of plants or their parts.
    VS : Visual recording by observation of individual plant or their parts.

A. Table of Characteristics

S. No.	Characteristics	States	Notes	Example variety	Stage of Observation	Type of assessment
1	2	3	4	5	6	7
1. (*)	Bush/tree height (m) (Vigour)	Small (<1.5)	3	Nana	Fruiting	MG
		Medium (1.5-2.5)	5	Bhagawa		
		High (>2.5)	7	IC-556897		
2. (*) (+)	Bush/tree growth habit	Upright	3	Gulesha Red	Vegetative	VG
		Spreading	5	Bhagawa		
		Drooping	7	IC-0599597		
3.	Precocity (year after planting)	Early (<2)	3	Bhagawa	First flowering	VG
		Medium (2-3)	5	G-137		
		Late (>3)	7	IC -318720		

Contd...

S. No.	Characteristics	States	Notes	Example variety	Stage of Observation	Type of assessment
1	2	3	4	5	6	7
4. (*)	Shoot thorniness (number of thorns per metre shoot length)	Absent	1	--	Fruiting	MG
		Less (<5)	3	Nana		
		Medium (5-10)	5	Bhagawa, Ganesh		
		High (>10)	9	IC-0599601		
5.	Bush/tree foliage density	Sparse	3	IC-444199	Fruiting	VG
		Medium	5	Bhagawa		
		Dense	7	IC-318705		
6.	Leaf blade length (cm)	Short (<2)	3	Nana	Fruiting	MS
		Medium (2-5)	5	P-13, Phule Arakta		
		Long (>5)	7	Patna-5, Bedana Suni		
7.	Leaf blade width (cm)	Narrow (<1)	3	Nana	Fruiting	MS
		Medium (1-2)	5	Dholka, Kasuri, Bhagawa		
		Broad (>2)	7	-		
8. (+)	Leaf blade shape	Elliptic lanceolate	3	Patana-5	Fruiting	VG
		Lanceolate	5	Malta		
		Broad elliptic	7	Ruby		
9. (+)	Leaf apex shape	Acute	3	G-137	Fruiting	VG
		Obtuse	5	KRS		
		Rounded	7	IC-524029		
10.	Petiole length (mm)	Short (<4)	3	IC-318716	Fruiting	MS
		Medium (4-6)	5	Kandhari, Damini		
		Long (>6)	7	Patna-5		
11.	Petiole anthocyanin colouration (% part covered)	Low (<25)	3	Achikdona	Fruiting	VG
		Medium (25-50)	5	Ruby, Patna-5		
		High (>50)	7	Malta, Kalisirin		

Contd...

S. No.	Characteristics	States	Notes	Example variety	Stage of Observation	Type of assessment
1	2	3	4	5	6	7
12. (+)	Calyx length (mm)	Short (<20)	3	Nana	Flowering	MS
		Medium (20- 40)	5	Damini, Nimali		
		Long(>40)	7	Kalisirin		
13. (+)	Calyx width (mm)	Narrow (<10)	3	Nana	Flowering	MS
		Medium(10-15)	5	Phule Arakta, Bhagawa		
		Broad (>15)	7	KRS		
14. (*)	Calyx colour	Yellow	3	Kabuli Yellow	Flowering	VG
		Orange	5	Ganesh		
		Red	7	Gulesha Red, Phule Arakta, Mridula		
		Other	9	-		
15.	Corolla colour	White	3	Kabuli Yellow	Flowering	VG
		Orange	5	Yercaud, Ganesh		
		Red	7	Bhagawa, Mridula		
		Other	9	-		
16.	Corolla type	Single	1	Nana	Flowering	VS
		Double	9	IC-444197		
17. (+)	Petal length (mm)	Short (<15)	3	Nana	Flowering	MS
		Medium (15-25)	5	Damini, Gulesha Red		
		Long (>25)	7	Jodhpur Red		
18. (+)	Petal width (mm)	Narrow (<10)	3	Nana	Flowering	MS
		Medium (10- 20)	5	Kalisirin		
		Broad (>20)	7	Jodhpur Red		
19. (*) (+)	Fruit length (cm)	Short (<6.0)	3	Nana	Fruiting	MS
		Medium (6.0-8.0)	5	Yercaud		
		Long (>8.0)	7	Ganesh		

Contd...

S. No.	Characteristics	States	Notes	Example variety	Stage of Observation	Type of assessment
1	2	3	4	5	6	7
20. (*) (+)	Fruit diameter (cm)	Small (<5.0)	3	Nana	Fruiting	MS
		Medium (5.0-7.0)	5	Dholka		
		Large (>7.0)	7	Ganesh		
21. (*) (+)	Fruit shape (Ratio of longitudinal and lateral axes)	Round (1.0-1.1)	1	P-26	Fruiting	MS
		Ovate (1.1- 1.2)	3	Bhagawa		
		Oval (1.2- 1.3)	5	-		
		Elliptical (>1.3)	7	Bedana Suni		
22. (*)	Fruit colour	Yellow	3	Kabuli Yellow	Fruiting	VG
		Red	5	Bhagawa		
		Deep red	7	Mridula, Phule Arakta		
		Other	9	Ganesh		
23.	Rind thickness (mm)	Thin (<3.0)	3	Nana	Fruiting	MS
		Medium (3.0-5.0)	5	Muscat, Bhagawa		
		Thick (>5.0)	7	Patna-5		
24. (*) (+)	Nipple or fin	Absent	1	Bhagawa	Fruiting	VG
		Present	9	Ruby, Ganesh, Mridula		
25. (+)	Crown length (mm)	Short (<15)	3	Nana, Kabuli Yellow	Fruiting	VS
		Medium (15-25)	5	P-23, Muscat		
		Long (>25)	7	-		
26. (+)	Crown neck	Absent	1	P-23	Fruiting	VG
		Present	9	Ganesh, Mridula		

Contd...

S. No.	Characteristics	States	Notes	Example variety	Stage of Observation	Type of
1	2	3	4	5	6	7
27. (*)	Aril colour	White	1	P-16	Fruiting	VG
		Light yellow	2	Kabuli Yellow		
		Light pink	3	Dholka		
		Pink	5	Ganesh		
		Red	7	Bhagawa		
		Dark red	8	Phule Arkta, Mridula		
		Other	9	-		
28. (+)	Aril length (mm)	Short (<10)	3	Bhagawa, Bedana Thinskin	Fruiting	MS
		Medium (10-15)	5	Ganesh, Patna- 5		
		Long (>15)	7			
29. (+)	Aril width (mm)	Narrow (<5)	3	Nana	Fruiting	MS
		Medium (5-7.5)	5	Ganesh, P-26		
		Broad (>7.5)	7	-		
30. (*)	Seed hardiness (Newton)	Soft (<25)	3	Bhagawa, Ganesh, Jyoti	Fruiting	MS
		Medium (25-45)	5	Kandhari, Ruby		
		Hard (>45)	7	Kabuli Yellow		
31. (+)	Seed length (mm)	Short (<6)	3	Nana	Fruiting	MS
		Medium (6-10)	5	Kalpitiya, Mridula		
		Long (>10)	7	-		
32. (+)	Seed width (mm)	Narrow (<2.5)	3	Nana, Kandhari	Fruiting	MS
		Medium (2.5-5.0)	5	Bedana Suni, Jyoti		
		Broad (>5.0)	7	-		

Contd...

S. No.	Characteristics	States	Notes	Example variety	Stage of Observation	Type of
1	2	3	4	5	6	7
33. (*)	Fruit maturity (days after anthesis)	Early (<130)	3	Nana	Fruiting	MG
		Medium (130-175)	5	Mridula, Ganesh		
		Late (>175)	7	Bhagawa		
34.	Total Soluble Solids (TSS) 0Brix	Low (<12.5)	3	Nana	Fruiting	MS
		Medium (12.5-16)	5	Bhagawa, Mridula		
		High (>16)	7	IC-318705		
35.	Acidity (%)	Low (<0.5)	3	Ganesh, Bhagawa, Mridula, Jyoti	Fruiting	MS
		Medium (0.5-1.25)	5	-		
		High (>1.25)	7	IC-318779, Nana		
36.	Fruit juiciness (%)	Low (<50)	3	Yercaud	Fruiting	MS
		Medium (50-60)	5	Ganesh, Jyoti, P-13		
		High (>60)	7	Mridula		

G. Special Test Characteristics

Under Rule 29 (b) of PPV&FR Rules, 2003, in case of failure of DUS test to establish the requirements of distinctiveness, the candidate variety shall, on request of applicant, be evaluated for under mentioned characteristics as special test.

S. No.	Characteristics	States	Notes	Example variety	Stage of Observation	Type of assessment
1	2	3	4	5	6	7
1.	Tolerance against abiotic stresses	Low	3	-	One year old plant	MS
		Medium	5	-		
		High	7	-		
2.	Tolerance against biotic stresses	Low	3	-	One year old plant	MS
		Medium	5	-		
		High	7	-		

H. Explanation for the Table of Characteristics

Characteristic 2: Bush/tree growth habit

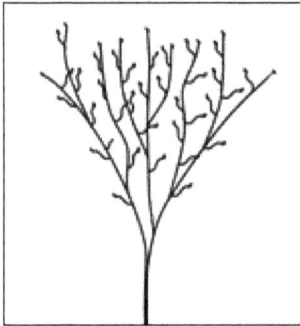 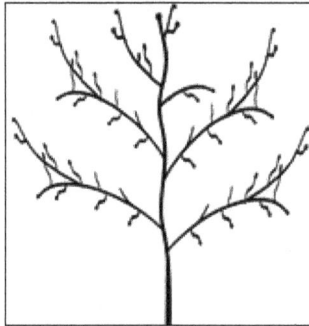 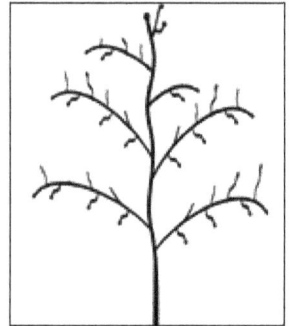

| **3**
Upright | **5**
Spreading | **7**
Drooping |

Characteristic 8 : Leaf blade shape

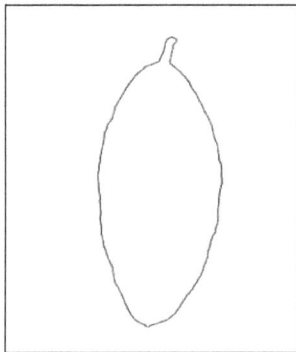 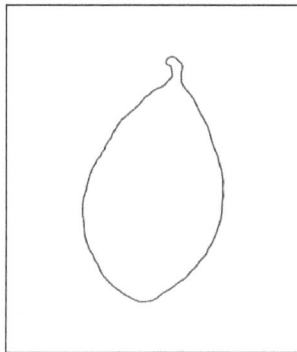 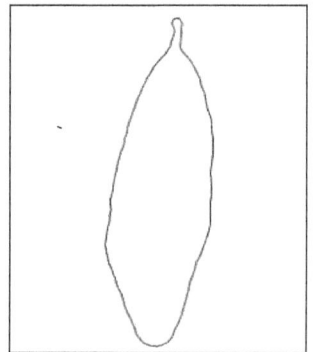

| **3**
Elliptic lanceolate | **5**
Lanceolate | **7**
Broad elliptic |

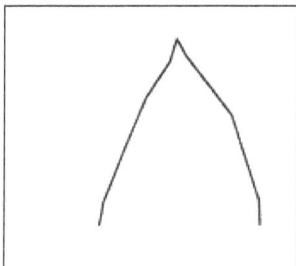 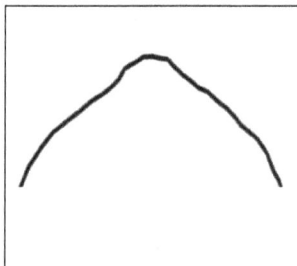 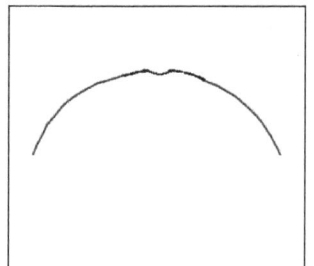

| **3**
Acute | **5**
Obtuse | **7**
Rounded |

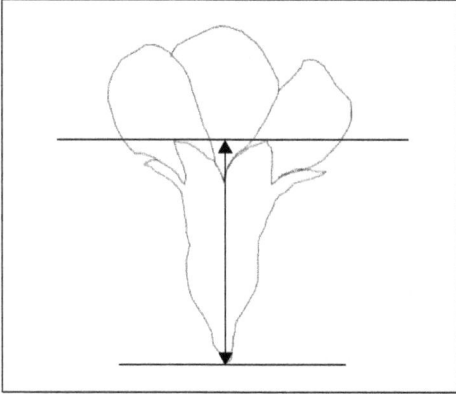

Characteristic 12: Calyx length

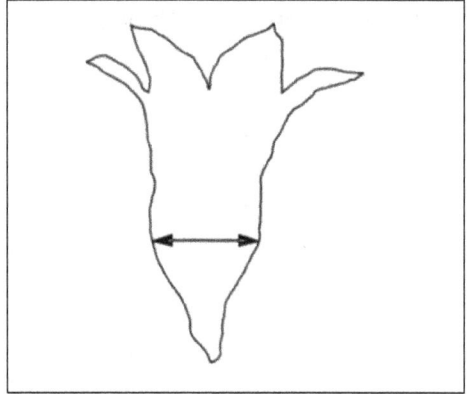

Characteristic 13: Calyx width

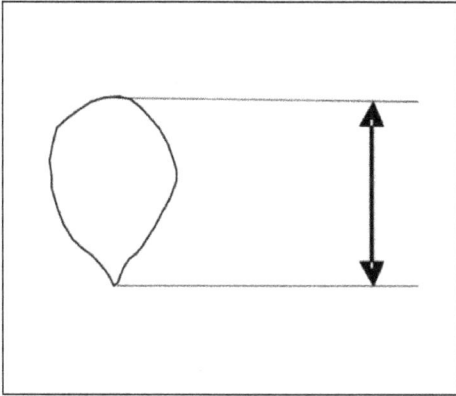

Characteristic 17: Petal length

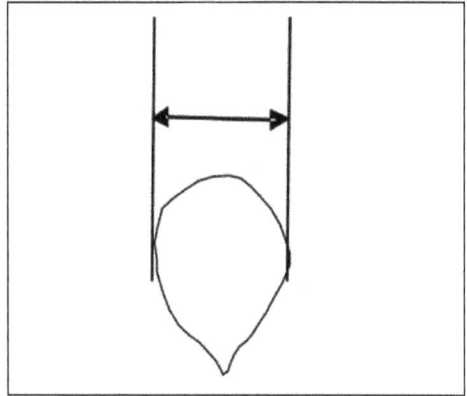

Characteristic 18: Petal width

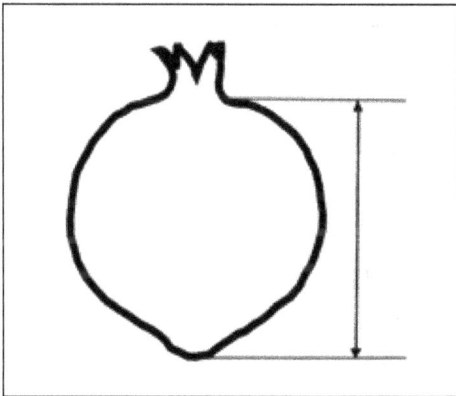

Characteristic 19: Fruit length

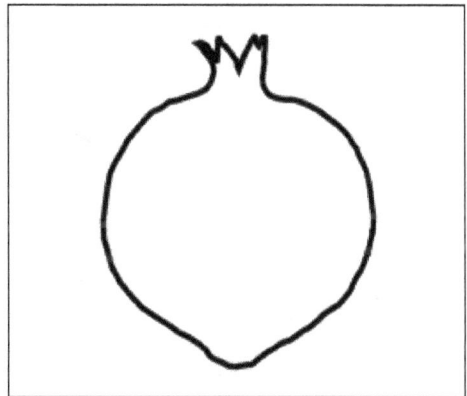

Characteristic 20: Fruit diameter

Characteristic 21: Fruit shape

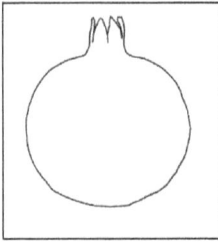 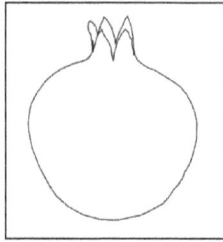 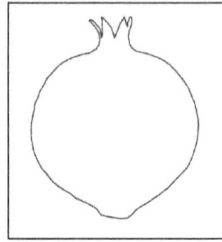 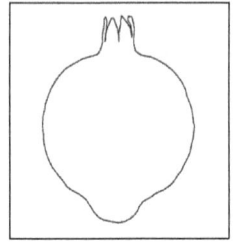

1	3	5	7
Round	Ovate	Oval	Elliptical

Characteristic 24: Presence of nipple on fruit

 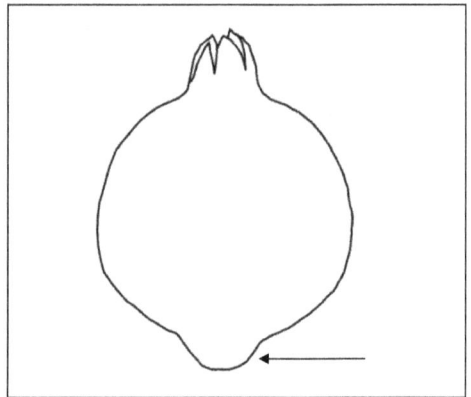

Absent	Present

Characteristic 25: Fruit crown length

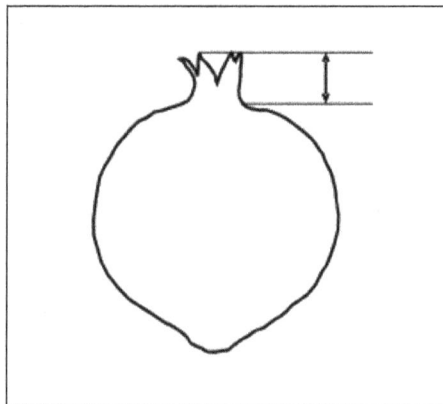

Characteristic 26: Presence of crown neck on fruit

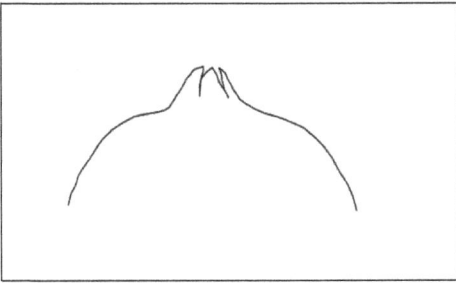

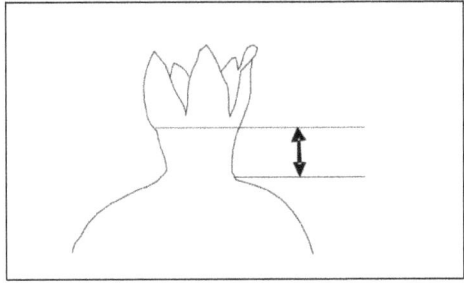

1
Absent

9
Present

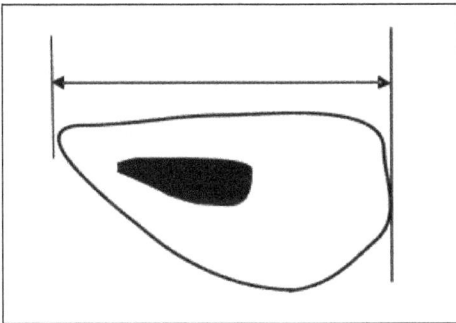

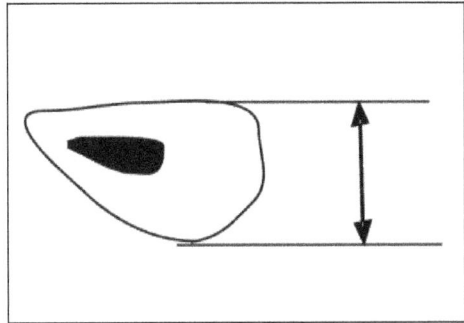

Characteristic 28: Aril length

Characteristic 29: Aril width

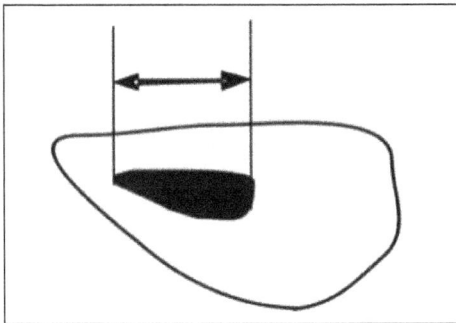

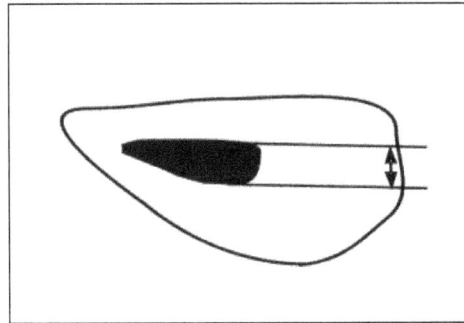

Characteristic 31: Seed length

Characteristic 32: Seed width

I. DUS testing centres

Nodal DUS test centre	Other DUS test centre
National Research Centre on Pomegranate, NH-65, Kegaon, Solapur-413 255 (Maharashtra).	Central Arid Zo ne Research Institute, Jodhpur-342 003 (Rajasthan).

6

Agrobacterium Mediated Transformation of Pomegranate cv. Bhagawa with *Pflp* Gene for Bacterial Blight Resistance

T.R. Usharani, H.D. Sowmya and Sukhada Mohandas

ICAR-Indian Institute of Horticultural Research, Bengaluru, Karnataka-560 089

Introduction

Pomegranate (*Punica granatum* L.) is a high valued fruit-bearing deciduous shrub belonging to family *Lythraceae*. It is being grown since ancient times for its fruit, ornamental and medical purposes. It is indigenous to Iran and is cultivated extensively in Spain, Morocco, Egypt, Iran, Afghanistan, Arabia and other Mediterranean countries. In India, pomegranate is commercially cultivated in Maharashtra, Karnataka and small scale plantations are seen in Gujarat, Rajasthan, Tamil Nadu, Andhra Pradesh, Punjab and Haryana (Chadha, 2001). The pomegranate cv 'Bhagwa' is the most important cultivar grown in this belt occupying 80 percent of area under cultivation (Meshram *et al.*, 2012). Bacterial blight caused by *Xanthomonas axonopodis* pv. *puniceae* is a serious disease devastating pomegranate fruit production and export earning since its outbreak in the year 2000 (Sharma *et al.*, 2013). The disease was first reported by Hingorani and Mehta (1952) from Delhi, India. The losses may extend up to 60-80% in unmanaged orchards under epidemic conditions with 'Bhagwa' revealing maximum susceptibility (Chand and Kishun 1991; Sharma *et al.*, 2013). While chemical control and biocontrol measures are already in place (Raghuwanshi *et al.*, 2013; Poovarasan *et al.*, 2013), the need of the hour is to find long term solution to the problem and breed resistant varieties. Despite their nutritional importance and health benefits (Gil *et al.*, 2000), there is very little research effort has been directed towards its improvement. All conventional ways and means of controlling this disease have been failed. Transgenic approaches will offer the potential for direct transfer of these genes into current cultivars and developing source of resistance for utilization in further breeding programmes.

Potential Strategies for Developing Pomegrante Cultivars Resistant to Bacterial Blight

Diseases in plants caused by bacteria are difficult to control. The most effective, economical and environmentally friendly approach to control disease is the use of resistant cultivars in breeding programme. However the crop's long life cycle and a heterozygous genome structure conserved by out-crossing make breeding slow and expensive. To circumvent the problem there are options to use several antibacterial protein coding genes and bactericidal proteins through biotechnology tools and develop transgenics as best alternative. Bacterial blight is controlled by genetic resistance i.e., the presence of Resistance genes (R gens) have been exploited to develop bacterial disease resistant plants in few crops like rice (*Oryza sativa-XA 21*), tobacco (*Nicotiana tabacum*), tomato (*Lycopersicon esculentum* -R gene *Pto*) and pepper (R gene *Bs2*). However manipulating the host defense mechanism through novel tools like RNAi and inserting the antibacterial genes through codon usage and realizing high level expression using genetic engineering can be attempted. Low molecular weight antimicrobial peptides (AMPs), have bactericidal action on a selective range of gram negative and gram positive bacteria like magainin and cecropins, isolated from animals and plants have been proven effective to wide range of bacterial species including *Erwinia carotovora, Pseudomonas syringae, Ralstonia solanacearum* and *Xanthomonas campestris (Jan et al., 2010)*. Two potential transgenes for controlling BXW are those encoding hypersensitive response-assisting protein (*Hrap*) and plant ferredoxin-like protein (*Pflp*) from sweet pepper (*Capsicum annuum*) (Chen *et al.*, 2000). Both have been proven effective against related bacterial pathogens, such as *Erwinia, Pseudomonas, Ralstonia* and *Xanthomonas* spp., in many Crops. Attacins are another group of antibacterial proteins produced by *Hyalophora cecropia* pupae in response to bacterial infection. The expression in transgenic potato enhanced its resistance to bacterial infection by *E. carotovora* subsp. *atrospetica* (Arce *et al.*, 1999). Similarly the attacin A gene for resistance to *Xanthomonas citri* subsp. *Citri* in transgenic sweet citrus *Citrus sinensis* L. Osbeck was employed (Cardoso *et al.* 2010). Stable expression and phenotypic impact of *attacin E* transgene in orchard grown apple trees was studied (Borejsza-Wysocka *et al.*, 2010). Transcription activator–like (TAL) effectors-based gene editing produces disease-resistant rice (Li *et al.*, 2012).

The success story of field trial of *Xanthomonas* wilt disease-resistant bananas in East Africa is reported (by Tripathi *et al.*, 2014). With the currently available useful strategy to develop broad spectrum resistance to bacterial pathogens, manipulation of such defense genes would be more ideal.

Plant ferrodoxine like protein (*Pflp*) is among the potential transgenes that act via hyper sensitive response (HR) proven effective against *Xanthomonas* infecting banana (Tripathi *et al.*, 2014). *Pflp* is a member of protein Ferredoxin-I (Fd-I), involved in several important metabolic pathways such as photosynthesis, nitrate reduction and lipid synthesis (Curdt *et al.* 2000; Geigenberger *et al.* 2005). It plays an important role by producing active oxygen species (AOS) and by activating hypersensitive response (HR) characterized by rapid, localized cell death on encounter with a microbial pathogen (Dangl *et al.*, 1996; Goodman and Novacky, 1994). Cell death resulting from HR forms a physical barrier to prevent further spread of pathogen leading to the development of systemic acquired resistance (SAR). *Pflp* is reported to be a potential candidate for *Xanthomonas* resistance in many transgenic crops. The *Pflp* gene has shown resistance against various bacterial pathogens like *Erwinia*, *Pseudomonas*, *Ralstonia* and *Xanthomonas* in transgenic tobacco, tomato, orchids, calla lily, rice and banana (Huang *et al.* 2004, Tang *et al.* 2001; Liau *et al.* 2003; Yip *et al.* 2007; Tripathi *et al.*, 2012).

The development of successful transgenic also depends on the regeneration capability of the explants used in transformation studies. The standardization of Agrobacterium infectivity for efficient transfer and molecular analysis to confirm the transgenics contribute to the generation of stable transformants. An attempt in this direction at IIHR resulted in developing few transgenic lines with *pflp* gene using a cotyledon explants and based on the *Agrobacterium tumefaciens mediated transformation* technology. These transformed lines of Bhagwa cultivar have been validated via PCR assay and southern blot analysis. They have been tested for disease resistance under laboratory conditions. The most promising lines will be evaluated for efficacy against Xap in fields. The transgenic lines will also be tested for environmental and food safety in compliance to our country's regulations.

Conclusion

Bacterial diseases are a real challenge to be effectively managed either by chemicals or biological means. The use of resistance source can provide the best way to combat disease. However if there are no germplasm exhibiting resistance. The rDNA technology can offer a promising, cost effective and timely approach. Plants trigger a varied defence mechanism in response to attack of bacterial pathogens. And as of today these defence genes are major target for developing transgenics. Selecting highly efficient antibacterial genes is need of an hour for developing resistant plants.

7

High Density Planting: An Approach for Enhancing Productivity in Pomegranate (*Punica granatum* L.)

Ram Chandra, Prativa Sahu, N.V. Singh,
K. Dhinesh Babu and R.K. Pal

ICAR-National Research Center on Pomegranate, Solapur– 413 255, Maharashtra
E-mail: rcpnrc@yahoo.co.in

Introduction

Pomegranate is highly suitable under tropical and subtropical regions of India due to its hardy nature, prolific bearing habit and more profitability. High density planting (HDP) system in pomegranate may be an option for farmers to improve the productivity per unit area. In fact, high density planting is a concept to enhance productivity with small land holdings by accommodating more number of plants per unit area without deteriorating the soil health. There is consistently growing trend to practice HDP among the farming community, especially in arid and semi-arid regions of Deccan plateau for various fruit crops including pomegranate. The HDP improves orchard efficiency, promotes precocity, supports mechanization and other efficient orchard management practices and ensures higher yield with better fruit quality. However, topographical conditions of land, low soil fertility, poor water quality and water scarcity in Deccan plateau do not support high density planting in pomegranate on sustainable basis without resource augmentation and technological backup. However, with organic and precision farming techniques, profitability of pomegranate could be enhanced with moderate plant population over normal planting system. Though, very little information on this aspect in pomegranate is available globally including India, systematic effort is needed in years to come with modern Research and Development (R&D) approaches. Increasing pressure on land owing to diversion of orchard land to various other obvious reasons as well as rising energy and land-costs, together with mounting demand for fruits have made it imperative to achieve higher productivity of pomegranate from limited space.However, one should be very conscious with HDP in pomegranate as closer spacing may have negative effect on sustainability at farmer's field if the complete package of HDP are not understood and followed judiciously.

Advantages of High Density Planting

- Efficient utilization of natural resources.
- Low labour cost and better orchard management.
- The ratio of fruiting shoots to supporting shoots is higher.
- Possible to accommodate more plants per unit area leading to higher income and quality fruit production.

Concept of High Density Planting

There is a shift in farmer's perception from production to productivity and profitability which can be achieved through high density planting. Recently, there is a trend to plant fruit trees at closer spacing leading to establishment of high density orchard. Higher and quality production is achieved from densely planted orchards through judicious canopy management and adoption of suitable high tech horticultural practices.

Establishment of Orchard

In order to utilize adequate sunlight equally on either side of the rows and save the crop from sunburn of the fruits and better aeration, north-south direction is considered ideal. As far as planting system is concerned square or rectangular planting systems can be followed in pomegranate. Square system is more suitable for wider spacing with vigorously growing varieties. However, a rectangular system is most suitable and popular in pomegranate. In fact, planting distance will depend upon soil type, climatic conditions and varieties.Pits of 1m x 1m x 1m or 0.75m x 0.75m x 0.75m are dug at a spacing of 4.5m x 3.0 m, especially for cv. Bhagwa. This can accommodate about 740 trees/ha. However under HDP system some concepts like prevention of upright growth, encourage horizontal branches and space small laterals along the strong limbs have to be followed. However, these could be achieved by dwarf and semi-dwarf varieties, training and pruning, mechanical devices and chemicals.HDP may be avoided in steep sloppy land and unfertile soil. Raised bed planting system for moderate to high density planting system, especially in low drained and salty soils should be followed. This system can drain excess salts available in the soil and improve aeration into the root zone. Need based inter crops may be grown in the orchard to get additional income during non fruiting or gestation period (within 2 years after planting). Intercropping, green manuring, hoeing, organic manure application, staking, pinching, soil amendment *etc.* should be done as per the need. All possible efforts are required to improve soil condition using beneficial microorganisms (biofertilizer) or organic manures (neem cake, FYM, vermicompost, leaf mould *etc.*). Selection of intercrops should be based on soil and environmental condition, flow of finance and market demand, availability of labourers *etc.* Keep better drainage facility within the plots and blocks.

Canopy Management

Training and pruning are the major cultural operations followed for developing appropriate plant canopy. Untrained and unpruned pomegranate trees become huge and unmanageable after a couple years of growth. Consequently, the bearing area is reduced and overall productivity of the tree is also reduced. Thus training and pruning are needed to increase the yield of quality fruit following appropriate principles of pruning. In fact, pruning begins at an early stage of plant growth to develop desired well spaced scaffold branches to form the strong framework. There are many ways to train and prune different fruit crops and no single method is right for all situations and needs. Therefore, systematic attempts are made meticulously in the first and second year of planting itself. Canopy architecture has significant impact on fruit production which is determined by the number, length and orientation of branches and shoots. In any fruit crop, for optimum fruiting and quality fruit production, the canopy management of the tree is prerequisite that deals with the development and maintenance of their structure in relation to the size and shape. The basic idea of canopy management is to manipulate the tree vigour and use maximum available light and temperature to increase productivity, fruit quality and also to minimize the adverse effects of the weather. Pomegranate is a light loving plant thus enough light should penetrate into the tree canopy for better flowering, fruiting and quality fruit production. The second factor is the planting system which includes both tree arrangement in the orchard (planting distance, row orientation) and manipulation of tree shape and height. Restricting the build-up of micro-climate congenial for the development of diseases and insect pests and convenience in carrying out the cultural operations are also important considerations. However, balance between vegetative and reproductive growth must be maintained giving emphasis to have less wood and more fruit on plant. Even there should be balance between shoot and root growth for better fruiting.

Moderate thinning and opening of the crowns of the trees planted at closer spacing in hedge rows found to increase light penetration in some fruit crops and similar system has to be followed in pomegranate for improving the productivity. The leaves growing in full sun, take advantage of maximum photosynthesis. Use of growth retardants, training and pruning systems and dwarf- semi dwarf cultivars in multi row bed system of planting may enhance light interception and distribution. Thin vertical canopies receive light exposure from both side of canopy. Smaller, more intensively managed trees will mean that more fruit will be produced on exposed parts of the tree. This is likely to affect fruit development and quality. Therefore, these aspects need to be critically considered in pomegranate. Short statured varieties could be exploited for enhancing fruit yield per unit area. In other fruit crops upright growing cultivars, make more efficient use of the orchard space and with simple adjustments to canopy architecture can lead to extremely high orchard yields. Several varieties of pomegranate having erect growth habit are available and they can be utilized in breeding programme for development of varieties suitable for desired architecture that can capture and utilize sunlight efficiently for quality fruit production.

Table 7.1: Categories of HDP, Spacing, Cultivars and Plant Vigour in Pomegranate

Categories of HDP	Number of plants (per hectare)	Spacing (m)	Cultivars	Plant height (m)
Low	<500	4-6	Ganesh, Bhagwa	2.25-3.0
Moderate	500-800	2.5-4.5	Ganesh, Bhagwa	2.0-2.75
High	800-1250	1.5-4.5	Bhagwa,Mridula	1.5-2.5
Ultra high density	1250-2000	1.5-4.0	Bhagwa	1.25-2.0

Even though a small canopy with a high number of well-illuminated leaves is efficient in photosynthesis but it is very poor in light interception, which leads to low potential yield per hectare. Light interception could be improved by increasing tree density. As a result, yield per hectare is maximized. Depending on the planting system, tree density under different system of planting (Table 7.1) in pomegranate will depend on tree vigour, spacing, soil type *etc.* Tree arrangement in HDP system must have sufficient alleyways for movement of farm machinery. In pomegranate, single hedge row with enough alley space can be practiced.

The tree growth of pomegranate in high density planting remains similar as in low density up to 3 years. The training begins when the tree is first planted and continues up to 3 years. However, proper tree forms, branch angle and limb spacing itself aids in growth control. Both single and multitraining system (3-4 main stem) may be followed. In single stem training system, well developed single stem saplings of about one year old should be planted. Cut the plants back to 35-45 cm at planting or after1-2 months of planting and allow to develop 3-4 symmetrically spaced scaffold limbs, the lowest at least 25-30 cm above the ground. Shorten scaffold limbs to 50% of their length. Remove interfering branches and sprouts leaving 2 or 3 shoots per scaffold branch (limb).Thus, a total 6-9 strong laterals (thick ones) will develop. Prune to produce stocky, compact framework in the first 2 years of growth. In multi stem training system, 3-4 healthy and strong suckers may be allowed in first year after planting and other suckers arising from the ground level should be removed regularly. Cut back selected shoots at 45-60 cm and allowed to grow. Select 2 healthy branches on each primary branch. Finally a total of 6-8 strong secondary branches will develop. Within 2 years after planting, such branches will develop desired canopy. Fruiting may be taken from third year after planting. Continue shoot pruning each year to maintain the tree shape and size as per training system.

Mechanization

Another component in HDP is the system automation which contributes to high production efficiency. One of the important farm operations that can be automated is irrigation and fertigation vis-à-vis indiscriminate mechanical pruning. Apart from this, mechanization should be done for regular spaying of chemicals for plant protection and other inter-cultural operations.

Nutrient and Water Management

Recommended doses of manures and fertilizers should be applied at appropriate stages of crop growth. Judicious irrigation should be given through drip irrigation system at regular intervals depending upon climate, soil and crop growth and its fruiting stage.

Plant Protection

Among the insect pests, fruit borer, thrips, aphids, shot hole borer, stem borer, fruit sucking moth are more important and their timely management should be done. However, bacterial blight, wilt, leaf and fruit spot diseases have been reported to cause high economic loss and these diseases should be controlled as soon as they appear.

Conclusion

High-density cultivation is the new mantra for raising the productivity of high-value commercial crops. Certain important strategies have been identified for enhancing horticulture development in India in order to be competitive in the world market. They involve adoption of modern, innovative and hi-tech methods. One such strategy is the HDP which includes adoption of appropriate plant density, canopy management, quality planting material and support management system with appropriate inputs. HDP technology results in maximization of unit area yield and availability of the fruits in the market as per need which fetch better price. Apart from high yields, the added advantage of HDP is to improve orchard efficiency. It promotes precocity, supports mechanization and other efficient orchard management practices and ensures better fruit quality. In view of the popularity of the high density technology and likely benefits, it is now high time to encourage further adoption of this technology by the pomegranate growers of the country. Availability of institutional credit for adoption of this technology would definitely popularize it further among the farmers. In the long run with HDP not only pomegranate productivity is expected to improve on sustainable basis but livelihood security of small and marginal farmers would be ensured in rural areas.

8

Pomegranate Cultivation in Semiarid Areas of Gujarat (*Punica granatum* L.)

S.S. Hiwale

Central Horticultural Experiment Station, CIAH –ICAR Vejalpur-389 001, Gujarat

Pomegranate (*Punica granatum* L.) is one of the important semiarid fruits cultivated in over 1.93 lakh hectare areas in India. Its cultivation is possible even on marginal degraded lands. Apart from this its ability to with stand salinity in soil and water to some extent made this crop to emerge as a hardy fruit crop. It is commercially grown for its sweet-acidic fruits, which provide a cool refreshing juice, and is valued for its medicinal properties. Its popularity is also due to the ornamental nature of the plant especially due to bearing bright red flowers throughout the year. Its antioxidant properties are well known. The juice and seed contain large quantities of tannin and puniciic acid, which is essential in curing several diseases. The sour types can be used for making *Anardana* which is a used as condiment in northern India. The fruit can be successfully grown even under purely rain fed conditions in semiarid areas. In the recent years pomegranate cultivation has became an economically viable proposition. Large acreage covered under new varieties like Bhagwa, Arakta and Mridula. The work carried out at CHES, Vejalpur since past 20 years has lead to identification of technologies leading to successful cultivation in semiarid areas of Gujarat, where the area under the crop is increasing fast, 5800 hectares in 2010-11 from 4000 ha in 2008-09 Varietal improvement has led to revolution in pomegranate cultivation. The varieties Ganesh, Mridula, Arakta and Bhagwa are popular in Maharashtra. Bassein Seedless, Jyoti, Ganesh, Mridula, Arakta and Ruby in Karnataka, Dholka in Gujarat, Jalore Seedless, in Rajasthan, Kabul Red, Vellodu, Yercaud-1, CO-1 in Tamil Nadu.The area under Cv.Bhagwa is increasing faster owing to attractive red colour of fruit and arils and very good keeping quality.

Commercial pomegranate orchards are confined mainly to Mediterranean, and semi- arid regions of the world. It is hardy fruit, which can be grown successfully, even in low fertile soils. Although pomegranate of high quality can be grown only where there is a cool winter and a hot dry summer, the trees grow under wide range of climatic conditions. It can be grown from the plains to an elevation of about 1829 m. The tree can withstand frost but gets injured by temperatures below $11.1°C$.

In pomegranate ideal recommendation for spacing is 5x5m but with the mechanization of cultivation and adoption of drip irrigation system spacing adopted by the farmers is 14x14 feet, 12x12 feet, 15x10 feet 10x10 feet, 12x10 feet.

Bahar treatments like root pruning, root exposure, withholding water, defoliation of plants by chemical *etc.*, are practiced to induce moisture stress, so that the plants drop their leaves and the growth will be controlled. The main objective of these treatments is to regulate the crop by forcing tree to take rest and produce profuse flowering and fruiting during any one of the three *bahars*. Allowing a particular *bahar* or flowering depend on market rate of fruits and arrival of produce to the market, water availability to the plants, type of soil, insect pest menace and weather conditions during crop growth period. For *mrig bahar*, plant growth has to be suppressed during April-May by withholding water. This is generally followed in South India. At CHES, Vejalpur, work on pomegranate cv. Ganesh under rain fed condition revealed that it is possible to raise successful *hasta bahar* crop. Hand thinning was used to remove *mrig bahar* flowering. Comparative studies on *mrig* and *hasta Bahar* indicated the fruit set, retention and yield per plant were highest in *hasta bahar*. The fruit of the bahar are available when there is no glut of pomegranate in market and hence fetch remunerative prices. The quality of fruit in respect of skin color, T.S.S.was also superior. Also the incidence of pest and diseases is also least. Fruit borer infection in mrig bahar was recorded up to 18.83 per cent whereas in hasta bahar it was just 8.92 per cent.

Drip irrigation /fertigation is now being implemented by the government as well as by farmers. In case of flood irrigation during winter irrigation should be given at the interval of 12-15 days and during summer at the interval of 5-7 days depending on the bahar. However with the limitation of water availability now a day's drip irrigation/ fertigation is becoming popular and in Maharashtra.According to one of the estimates there is 98 percent increase in production with a saving of water to the tune of 45 percent. At M.P.K.V. Rahuri combined application of fertilizers by conventional method and through drip revealed that the yield obtained (11.88 t/ha) due to 100percent NPK recommended dose of solid soluble fertilizers through drip was significantly superior over 100 percent Recommended dose of conventional fertilizer and was at par with 70% N, 80% P and K and 70% NPK. Thus the maximum fertilizer saving (N, P and K) was 30 percent each compared to conventional fertilizer.

Even though pomegranate grows well in soils of low fertility, production can be increased by application of manures and fertilizers. Initial soil analysis is desirable for proper scheduling of fertilizer. The nutrition recommendation depends on fertility of the soil and also the age of the plant. For young plants (2-3 months old) application of 200 g Neem cake along with 3-4 kg FYM/plant is recommended. After 3 months, each plant may be given 250 g DAP along with 250 g Neem cake and 5 kg FYM. Again after 9 months, application of 500 g DAP, 250 g Potassium sulphate, 1 kg Neem cake and 10 kg of FYM/plant is recommended.

Under Rahuri conditions 1 ½ to 2 year old plants need 250 g N, 125 g P, 125 g K, 2 ½ to 3 year old plants need 500 g N, 125 g P, 125 g K, 4 ½ to 5 year old plants need 500 g N, 250 g P, 250 g K and 6-7 year old plants need 625 g N, 250 g P, 250 g K.

At IIHR, Bangalore, application of 500 g N + 250 g P + 125 g K/plant/year to Bassein seedless variety of pomegranate gave the highest yield of 8.2 kg fruits from 3 year old plant (first crop) and application of 500 g N, 250 g P and 250 g K/plant/year to Cv. Ganesh gave the highest yield of 15.67 kg (50 fruits/plant) during fourth year of cropping.

With the increasing importance of organic farming in India andrising threat to food security and natural resources, organic pomegranate production is attracting farmers. With the advent of green revolution indiscriminate use of chemical fertilizers, pesticides weedicides had adversely affected the soil fertility, productivity and quality of produce. Fruit retention and yield per plant were significantly influenced by various treatments. Maximum fruit set (96.50 fruits / plant), fruit retention (57 fruits / plant) and yield (10.75 kg / plant) were recorded in treatment, application of nitrogen fifty per cent through FYM, twenty five percent through castor cake and urea. Similar results were reported by Shinde (1977). Phadnis (1974) also recommended mixed doses of organic and inorganic fertilizers to obtain maximum fruiting and yield.

Biofertilizers are preparations containing live or latent cells of efficient strains of nitrogen fixing, phosphate solubilizing or celluolytic microorganism used for application to seed, soil or compost with the objective of increasing the number of such organism which will accelerate the process of nutrient availability to the plant.

Biofertilizers had no influence on vegetative parameters of pomegranate. *Azospirillium* @ 100g per plant resulted in higher N content of leaf as well as higher fruit set and retention, however, fruit weight and fruit size and yield was maximum in combined application of PSB+ Azospirillium @ 50g per plant. Use of biofertilizers like Azotobactor has reduced the requirement of fertilizers. Application of 250 g *Azotobactor* culture with 100 g N has given the same effect as application of 300 g

N per plant, i.e., saves 200 g N per plant. Nutrient analysis of leaf samples collected one month after fertilizer application showed significant influence in respect of phosphorus and potassium whereas, no significant differences in respect of nitrogen content in various treatments. Maximum fruit set (144.75 fruits / plant) and fruit retention (83 fruits / plant) were recorded in *Azospirillium* 100g/ plant and yield (11.08 kg / plant) in 50g *Azospirillium* + phosphate solubilizing bacteria culture.

High-density orchard results in early bearing helping in minimizing weed problem. However, the productive life of orchard is reduced due to dense canopies, which will not allow sunlight penetration resulting in slow decline. This problem can be overcome by removing plant in alternate row. The other methods, which can be used in overcoming this problem, are training and selective pruning. Productivity of almost all the fruits in India is low as compared to other fruit growing countries of the world. To overcome this problem in pomegranate high-density orchard in cv. Ganesh has been standardized under Maharashtra condition. To obtain higher production per unit area maximum number of plants were accommodated per hectare. High destiny plantation at 5 x 2.2 m under semi arid condition resulted in 2.5 times higher yield as compared to normal spacing of 5x 5 m in 6 to 7 year old orchard.

The systematic work on this aspect was initiated at CHES Vejalpur where manual pruning was carried out. Pruning in pomegranate was attempted to initiate new growth on 5 year old plants. The plants were pruned in the second half of May. Two types of pruning i.e. heavy (removing around 5-7 kg fresh wt / plant and light (removing 2-3 kg of fresh wood/plant) was attempted. Under rainfed conditions. Sprouting started after 3-4 weeks of pruning. The subsequent flowering was however was delayed. Though fruit set was significantly higher in control (No pruning), fruit retention was higher in heavy pruning giving ultimately higher yield per plant. There was an increase of about 7.36 q/ha (16.95%) in yield, giving an additional income of Rs. 3680/ha @ Rs 500/q. Similarly fruit weight and fruit diameter showed significant increase, which might have resulted in increased yield per plant at CHES Vejalpur, under semiarid, rainfed conditions. Fruit weight showed maximum increase over control (79.92 percent) when 25 fruits/plant were retained; it was 69.08 percent with 50 fruit and 41.60 percent with 75 fruits per plant. However, there was no change in percent juice content and acidity of fruits. TSS content was maximum when less number of fruits/ tree were retained (17.08⁰ Brix), showing a decreasing trend as number of fruits per plant increased. It was observed that keeping 75 fruits per plant produced maximum yield per plant (11.41 kg) compared to control (10.24 kg). Thinning fruit to 25 numbers though resulted in reduction in yield by 54.98 percent and 12.56 percent in thinning to 50 fruits. Fruit thinning produced good size fruit of 'B' grade, fetching better prices in the market. Overall keeping 50 fruits per plant were found to be economically viable preposition to the farmers as the economic returns were highest (Rs.10670/-) compared to control (Rs.4692/-).

The old orchards become non-productive due to incidence of pest and diseases and neglect. In India once the orchard becomes old its productivity goes down, and it is general tendency of the farmers to neglect the orchard. The fruit set in the initial year was reduced in the first year of pruning (19 no/plant), which surpassed the control in second year itself (101 no/plant) in plants pruned to 30 cm from ground level. It outperformed control in the years to follow. Similar trends were recorded in respect of no. of fruit retained / plant. Maximum no of fruit were retained in plants pruned to 30 cm from ground level. Yield kg/plant was reduced in the first year after pruning to 1.14 kg (81.18% reduction) in treatment pruning to 30 cm from ground level. The yield/ plant however has surpassed control in second year itself and the trend continued thereafter. The increase was to the tune of 16.96%, 32.94% and 26.10% in 2nd, 3rd and 4th year of pruning to 30-cm.level.

Root distribution acts as guide for application of fertilizers as well as irrigation. Root distribution varies according to the type of soil, method of propagation. Root distribution pattern in air layers of pomegranate cv. Ganesh under semiarid rainfed conditions revealed that the root system is shallow in nature as below 60cm soil depth not much root activity was recorded. Maximum root activity on fresh and dry wt. basis was observed in0- 30 cm radial distance from tree trunk (54.17%), which was 31.12% in 30-60 cm radial distance and 2.12 % in 60-90 cm radial distance. However soil depth wise it was maximum at 30-60 cm soil depth and 42.69% at 0-30cm depth. It was least (3.74%) at 60-90 cm depth.

9

Effect of Mulches on Yield, Quality and WUE of Pomegranate (*Punica granatum* L.)

D.T. Meshram, Ashis Maity, N.V. Singh, Ram Chandra and R.K. Pal

ICAR-National Research Center on Pomegranate, Solapur-413 255, Maharashtra
E-mail: gomesh1970@rediffmail.com

In pomegranate growing regions of India, water is a scarce resource and its efficient use has to be prioritized. Regular water supply through irrigation system is of paramount importance for sustainable production of pomegranate. In Maharashtra, pomegranate is predominately grown in the regions of Desh (*i.e.* Solapur, Ahmednagar, Kholapur, Pune, Satara and Sangali), Marathwada (*i.e.* Aurangabad, Osmanabad, Beed, Hingoli, Jalna, Latur, Nanded and Parbhani); Khandesh (*i.e.* Dhule, Jalgaon, Nandurbar, Pune and Nasik) and Vidharbha (*i.e.* Buldana, Yawatmal, Akola, Amravati, Wasim, Nagpur, Gadchiroli and Wardha). In these parts of Maharashtra water is scarce commodity and hence there is need to apply water judiciously as per the water requirement of the crop. The water requirement of pomegranate crop depends on age, season, location and management strategies.

A mulch is a material spread on the ground surface to protect a plant or plant roots. Natural mulches such as dry leaf, straw, dead leaves, sugarcane trash, paddy husk, paddy straw, safflower, wheat straw and plastic mulch (*i.e.* black, white, pervious, silver and black, red *etc.*) have been found very effective in conserving soil moisture for minimizing the evaporation losses and maintaining soil temperature. The LLDPE, HDPE and flexible PVC mulches are effective in reducing reference crop evapotranspiration (ETr) as crop coefficient values decreases by 10-30 % due to the 50-80% reduction in soil evaporation, evpotranspirtion, environmental stress coefficient.

S field experiment was carried out during 2011-2014 using one to three years old pomegranate trees at ICAR-NRCP, Solapur (*i.e.* North latitude 17^0 10", East longitude by 74^0 42"and 483.5 m msl) to quantify yield, quality and water use efficiency of pomegranate Bhagawa Cv. at 4.5 x 4.0 m spacing under influence of inorganic mulch. A split plot design of three different types of in-organic mulches (*i.e.* black and silver, black and pervious mulches) at different irrigation levels ranged from 0.10 to 0.60 *ETr with four replications and alternate days irrigation were evaluated for their interactive. The physical, chemical and hydrological properties of the soil profile of experimental site have been evaluated before and after experimentations.

The amount of water applied was minimum 0.10*ETr and maximum 0.60*ETr through drip irrigation for black and silver, black, pervious and control under split plot design. Crop coefficient (Kc) and wetted area (%) were estimated ranged from 0.18 -0.80 and 20 - 40 % for one to three years old pomegranate trees. Maximum plant height, flowers, branches, stem diameter and fruits was recorded in pervious in-organic mulch. The actual water applied in different in-organic mulches treatments 10-60 % is less than the actual water demand due to the reduced wet evaporative surface. The actual water requirement (WR) varied from 3 - 30 litres / day / tree during different phenological stages and 0.20 to 0.40 *ETr irrigation levels was observed best for inorganic mulch (*i.e.* pervious) for one to three years old pomegranate trees.

Another field experiment was conducted at ICAR-National Research Center on Pomegranate Research Farm, Solapur, India during late Hasta bahar. The details of experiment was laid out with two factors in split plot design with main the treatment of five irrigation levels (*i.e.* Factor A: 0.30 to 0.70 * ETr of four years old pomegranate tree) and sub-treatments as mulch (*i.e.* Factor B organic: M0-No mulch, M1- wheat, M2-safflower, M3-sugarcane baggas and Factor C inorganic: M0-No mulch, M1- black, M2- black and silver, M3-pervious). The electrical conductivity and residual sodium carbonate of the irrigation water used was 0.5 dSm^{-1} and 2.2 meq^{l-1}, respectively. The drip irrigation system consisted of plastic laterals of 16 mm diameter with on-line pressure compensating drippers at 60 cm distance away from trunk of the trees. The drippers had a discharge rate of 4.00 l h^{-1} under an operational pressure of 1.0 kgcm^{-2}. The irrigation through drip system was applied at alternate day for required time to deliver the calculated quantity of water based on atmospheric demand. The experiment was conducted on light texture soil with standard recommended dose of fertilizers and other management practices.

The data revealed that the yield attributing traits responded differently to different quantities of water applied through drip irrigation system having two laterals with four drippers at different irrigation levels during both the hasta bahar.

The influence of organic and inorganic with optimum irrigation on fruit yield is envisaged from the fact that the increment in yield attributing triats (fruit weight and number of fruits) in pomegranate to the tune of about 15-40 % were recorded as compared to no mulch and lower levels of irrigation.

Yield attributing traits were significantly higher in organic and inorganic mulches at 50 per cent irrigation level for four year old pomegranate cv. Bhagawa under micro-irrigation. Based on statistical analysis of vegetative and yield characteristics, the organic and inorganic mulches at 50 per cent irrigation level with alternate day irrigation frequency resulted in higher number of fruits per tree alongwith increase fruit weight with quality. Hence, water management and use of mulching ensure increased crop yield, high water use efficiency, high water saving, energy consumption and minimal weed problems. It is concluded from the present study that, organic and inorganic mulch is the better technological option for improving crop productivity and water use efficiency.

Advantages

1. Prevents direct evaporation of moisture from the soil
2. Plastic mulches provide barrier to soil pathogen
3. Maintain warm temperature during night time
4. Plastic mulch play role in soil solarisation and avoids weed problem in plant canopy
5. Water erosion is averted
6. Organic mulch adds organic matter to the soil

Disadvantages

1. Plastic mulch is costly to use in commercial production
2. Reptile movement and rodent activities are experienced
3. Environmental pollution in case of plastic mulching
4. Difficulty in machinery movement

10

Quality Planting Material Production in Pomegranate

N.V. Singh, K.D. Babu, Ram Chandra, J. Sharma, Prativa Sahu, D.T. Meshram and R.K. Pal

ICAR-NRC on Pomegranate, Solapur-413 255, Maharashtra
e-mail:nripendras72@gmail.com

Introduction

Quality planting material is a key to success for proper orchard establishment and optimum production in pomegranate. Traditional propagation methods need to be confluenced with modern propagation techniques and logistics to fulfill the ever increasing demand of elite planting material in required quantity. On an average the yearly estimated demand for pomegranate planting material is 5-6 millions and this demand is increasing with quite a good pace. This ever increasing demand of pomegranate can only be met when conventional and non conventional methods of pomegranate propagation can be exploited at commercial scale with need based modifications by involving modern propagation technologies. Traditionally, pomegranate is propagated through air layering in major pomegranate growing belts of Deccan Plateau. Hardwood cutting is another method which is prevalent in rest of the India and other regions of pomegranate cultivation. The ICAR-National Research Centre on Pomegranate (NRCP) has standardized hard wood cutting and sanitation protocol with high cutting success (65-70%) for pomegranate cv. Bhagwa. Recently, some of the Indian organizations including ICAR-NRCP have come up with tissue culture protocol for pomegranate but endeavor is still under progress to produce bio-hardened and better field performing pomegranate plants using tissue culture technique in an economic manner. The utilization of *in vitro* propagated plants should be made mandatory for expansion of pomegranate to non-traditional areas, so as to avoid spread of pathogens like *Xanthomonas axonopodis* pv. *punicae* to new areas through infected planting material. Bio-hardening of *in vitro* raised plants by utilizing plant beneficial microbes and placing their formulations in rhizosphere and phyllosphere of *in vitro* raised plants, ascertain the improved field performance by virtue of their improved morphological, physiological and biochemical functioning. However, there is an urgent need to develop package of practices and sanitation measures for *in vitro* raised pomegranate orchards to realize complete benefit of this technology. Recently, the ICAR-NRCP has initiated systematic efforts for standardization of grafting and budding techniques and research for identification of suitable rootstocks against various biotic and abiotic stresses are in progress.

Nursery Certification

A. Preparation and care for disease free certified pomegranate nurseries

1. The selected mother plants should be maintained by the ICAR-NRCP/other ICAR institutes/NBPGR/ State Agricultural Universities /other institute. Relevant morphological/ genetic/ molecular markers should be established for maintaining their varietal identity and purity.

2. The progeny orchards should be established from mother plants or using plants of mother block in different areas free from bacterial blight. This should be regularly monitored by team of experts from ICAR and SAUs of respective regions.

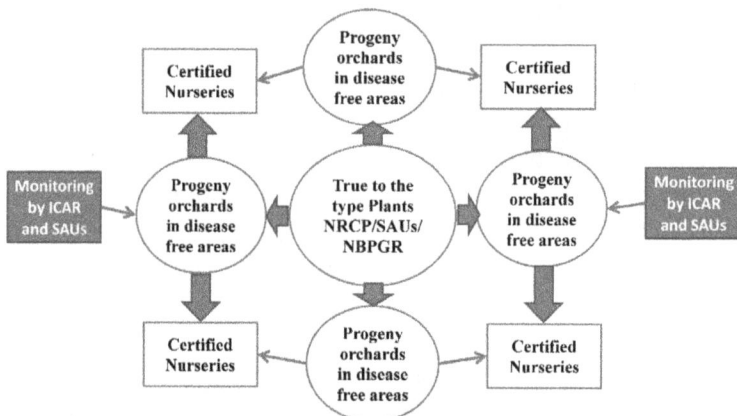

Steps to Produce Disease Free and Authentic Planting Material

4. Suspected propagating material has to be tested through diagnostic symptoms, ooze tests, microscopy, PCR based diagnostics and also isolation.

5. Mother plant/block used for air layering and hardwood cuttings should be sprayed and monitored regularly.

6. Treat the roots of air layered cuttings with copper oxychloride (COC) @3g/l to protect against soil borne diseases at the seedling stage and plant them in the standard size polyethylene bags filled with above potting mixture.

7. The nursery should be kept clean and sprayed regularly with fungicides, insecticides and if needed with bronopol or streptocycline.

8. Pruning tools – secateurs *etc.* if used should be sterilized after handling each plant with sodium hypochlorite (2.5%).

B. Recommendations for propagation through tissue culture

1. Mother plant with proven horticultural trait, obtained from reliable source and in healthy state should be used for excision of explants.

2. It should be ensured that the mother plants should be kept under protected structures/ insect proof shade net and monitored regularly for any visible symptoms of infection.

3. All the plant protection and phytosanitory measures should be regularly followed to mother plants healthy.

4. Shoot tips or meristem portion or axillary bud with nodal segments should be used as explants and other explants which follow indirect regeneration pathway should be avoided to reduce the chances of occurrence of somaclonal variations.

5. Number of multiplication cycles should be limited to 5-6 so as to reduce the chances of somaclonal variations.

6. Biohardening/biopriming of tissue culture raised plants should be carried out by using various plant beneficial microbes like arbuscular mycorrhizal fungi to improve the field performance of *in vitro* raised plants. At least 3-4 months of hardening should be done to reduce field mortality.

7. Clonal fidelity testing of *in vitro* raised plants should be carried out using molecular tools to detect variability if any.

Propagation

Most of the fruit crops are propagated through clonal or vegetative methods of propagation to avoid any variation in the planting material arising due to cross pollinated nature of fruit crops. Clonal propagation is also helpful in maintaining performance and genetic constitution of planting material exactly similar to the mother plant. In Deccan plateau air layering is the most common method of propagation and is practiced during June-August.

Hard Wood Cutting

Propagation of pomegranate by stem cuttings is a common practice in most of the pomegranate growing regions of the world. After observing cutting success form different methods of stem cutting, it can be concluded that pomegranate is not an easy-to-root plant by stem cuttings. Propagation by stem cutting requires formation of adventitious root system as potential shoot system is already there in the form of shoot buds. It is easy to observe initial sprouting in pomegranate cuttings to the tune of more than 85 percent but due to inability of some of the cuttings to form adventitious root system, the ultimate cutting success reduces considerably.

Type of Wood: Pomegranate is mainly propagated through hardwood stem cuttings because maturity of wood and amount of reserve carbohydrates play important role in rooting of cuttings. Six to eighteen months old shoots of previous season or even six months old shoots are also suitable for hard wood stem cuttings. The length and diameter of stem cuttings have an impact on cutting success and subsequent survival in the field. Generally, shoots with 6-12 mm diameter are used for preparation of cuttings. The length of the cutting is ideally kept in the range of 18-20 cm having 4 nodes. Hardwood lateral shoots, which usually flower and fruit heavily are not suitable for propagation, thus, non bearing shoots (erect type) may be selected for preparation of stem cuttings. In general, cuttings prepared from pruned wood of rest period register high cutting success probably because of accumulation of carbohydrates and other rooting factors in the shoots. An experiment was set up during 2011-12 and 2012-13 at ICAR-NRCP to identify the suitable length of hard wood cutting to optimize the amount of wood required for propagation without compromising the cutting success. At 120 days after planting there was no significant difference in the cutting success of 20 cm (76.33 %) and 15 cm (70.83 %) long cuttings but when cutting size was reduced to 10 cm, the success rate drastically came down to 45.33 per cent indicating the ideal hard wood cutting size as 15 cm.

Pretreatment of Cuttings: Treatment of cuttings with fungicides and antibiotic/ bacteriostatic compounds for sufficient duration helps in reduction of pathogen load. This pre-treatment also helps in making cuttings almost free from most of the surface inhabiting pathogens including *Xanthomonas axonopodis pv. punicae*. Complete submergence of cuttings in the luke warm (40 °C) aqueous solution of Bavistin™ (Carbendazim 50 % WP) @ 2 g/l along with Bactronol-100 (2-Bromo-2-nitro propane-1,3- diol) @ 0.5 g/l for 30 minutes, take care of most of the pathogens, copper fungicides are also effective in killing surface pathogen. This is followed by surface sterilization with sodium hypochlorite (NaOCl) or chlorinated water @2.0 % for 10 minutes this treatment is effective in minimizing pathogen load without hampering the cutting success. Carbendazim has also been reported to be effective in inducing roots in stem cuttings, however, the exact mechanism of stimulation of rooting by Carbendazim is not clear, it may be related to the auxin like activity of this chemical.

Root Promoting Growth Regulators: In pomegranate, stem cuttings are reported to be deficient in root promoting cofactors, auxin synergists and endogenous auxins level, thus exogenous application of auxins significantly influence the rhizogenesis. The root inducing growth regulators play critical role in hastening root formation, increasing percentage of rooting, promoting uniformity of rooting and enhancing quality and number of roots per cutting. In pomegranate, basal half portion of cuttings when treated with 2000-3000 mg/l IBA for 30 seconds to 5 minutes results into high cutting success.

Rooting Media: Rooting media also play important role in the root proliferation and further growth of plants raised by stem cuttings. It should possess certain desirable physical and chemical properties like optimum bulk density (30 % solid particles and 70 % space for air and water with bulk density between 0.2 - 0.5 g/cc), high water holding capacity, light in weight, optimum pH (5.5 to 6.5) and electrical conductivity. Cocopeat either alone or in combination with sand produced rooting success to the tune of more than 70 %.

Season: In pomegranate, the time of planting of stem cuttings in nursery and field conditions affects rooting and subsequent survival. In general, hard wood cutting taken immediately after rest period or pruned wood of rest period gives high cutting success. Hard wood cutting planted during June-August and November-February gave high cutting success under Solapur conditions.

Grafting

In many fruit crops identification of promising rootstocks and their utilization in the production process through grafting has significantly influenced the fruit industry. Presently, wilt is an emerging threat to the pomegranate industry in its major growing areas. Non- availability of wilt tolerant rootstocks is a major impediment for its mitigation. Research on suitability of rootstocks is the need of the hour to mitigate the challenges under climate resilient horticulture with special reference to increased soil salinity, drought, insect pests and diseases in commercial cultivars of pomegranate.

Wedge Grafting in Pomegranate: NRCP has standardized grafting technique in pomegranate and wedge grafting proved to be most successful method. 'Bhagwa' scion on different wild rootstocks gave about 90 per cent graft success at 45 days after grafting. For preparation of grafts by this method, 1-1½ years old rootstock are decapitated (headed back) at 25-30 cm above the ground level and beheaded rootstock is split to about 5cm deep through centre of the stem with a knife. Six to twelve months old scion of 15-20 cm length having 0.7-1.0 cm diameter is taken from a terminal shoot and it is prepared in wedge shape and inserted in the vertical split of the beheaded rootstock. Polythene tape or polythene strips of 200 gauze thickness are used for tying the grafts. The grafts are covered with 20-25cm long polythene tube and tied with thread at its lower portion (base). The grafts are watered and kept under shed to protect from direct sunlight. In general, scion sprouting starts between 8 and 12 days after grafting and after a few days of sprouting (12-15 days), the polythene tubes are removed. Wedge grafting during January under Solapur condition gave 90 % graft success. However, grafting can be done throughout the year under green house conditions.

Patch Budding

This method is recently standardized by ICAR-NRCP, Solapur and can be very effectively utilized after identification of promising rootstocks. It opens the avenue for *in situ* budding. More than 90 per cent success has been achieved when Bhagwa scion bud was budded on wild pomegranate root stocks during November to February under Solapur conditions. One year old rootstock and a patch bud of 20 mm x 10 mm containing some wood portion have been found ideal for patch budding. The patch of bark containing the bud is cut from the bud stick in the same manner in which the bark patch is removed from the rootstock. Two transverse cuts through the bark one above and other below the bud are made, then two vertical cuts are made on each side of the bud so that bark piece should be about 20 mm long x 10 mm wide. After the bud patch is removed from the bud stick it must be placed immediately on the rootstock. The patch from the bud stick should fit snugly at the top and bottom into the opening in the rootstock. Polythene tape or polythene strips of 200 gauze thickness is used for covering budded portion to enhance budding success by maintaining high moisture at the budded portion. The preliminary trials gave encouraging success to the tune of more than 90 per cent when Bhagwa scion was budded on wild rootstocks.

In Vitro Propagation

Conventionally, pomegranate is propagated through air layering and hardwood cuttings which do not ensure production of disease free planting material, moreover, success for these vary with season and maturity of the wood. The gaining popularity and up surging demand for quality planting material can't be fulfilled alone by conventional propagation methods and *in vitro* propagation has to be explored as an alternative tool to fulfill the demand of good quality disease free planting material in large number.

Shoot tips, axillary buds, hypocotyls (may lead to variations) and nodal segments have been used as explants in pomegranate. Callus can also be taken for regeneration and production of entire plant. Generally leaf segments, petals, hypocotyl and cotyledons are used for callusing. However, micropropagation through callusing may result abnormalities or sometimes creation of variability through enhanced possibility somaclonal variations, hence preferred only when other explant types are either not responsive or available *e.g.* coconut, datepalm, oilpalm. Preparative and explant plant pretreatment stage comprised of donor/mother plant selection and treatments, excision and treatments of explants (2-3 cm long and 20-25 days old nodal segments) with various antibiotic, fungicidal and surface sterilizing agents to avoid any microbial contamination and ensure good culture establishment Shoot proliferation of established culture occur through stimulation of branching, growth of side shoots and vertical splitting of elongated multinodal shoots. After 5-6 cycles of multiplication, sufficiently long shoots (3-4 cm) are transferred to rooting medium. Well rooted plantlets are hardened *in vitro* initially and then *ex vitro*.

Phenol Exudation and its Control (A Major Concern for Pomegranate *in vitro* Propagation)

Establishment of *in vitro* culture of several plant species, especially woody plants, is greatly hampered by the lethal browning of the explant and culture media. The various techniques employed to overcome the harmful effects of browning either attempts to neutralize or avoid the buildup of toxic substances in the media.

Choice of juvenile explants or new growth flushes during the active growth period, covering the mother plant with black cover a fortnight before taking explants or growing mother plants under controlled conditions, culture in liquid medium, inclusion of antioxidants in the culture media or soaking explants in water or solutions containing antioxidants prior to inoculation (antioxidants like ascorbic acid, citric acid), use of adsorbing agents such as activated charcoal (AC), polyvinylpyrrolidone (PVP), use of low salt media, sealing the cut ends with pre-sterilized paraffin wax and drying the explant under laminar airflow before inoculation, keeping the inoculated culture in darkness for one week and transfer of explant to fresh medium at short intervals (frequent subculturing).

Media Composition: Various mineral salt formulations have been used for *in vitro* culture of pomegranate, However, full strength mineral salts are not always optimum. Hence different media formulations may work better at different morphogenetic stages. Modifications with respect to different constituents like phytohormones, vitamins, macro and micro salts, sucrose, agar concentrations and other additives like PVP, AC, coconut milk, *etc.* are usually done in order to ensure better *in vitro* response.

Basal Media: Different media, *viz.* Murashige and Skoog Medium (MS), Gamborg's B5 and Woody Plant Medium (WPM) are used for *in vitro* propagation and among these MS and its modified forms are most commonly used for pomegranate. There are reports on utilizing WPM basal medium for rooting of microshoots in pomegranate.

Plant Growth Regulators: Most of the studies support the use of BAP, NAA and IBA as major growth regulators involved in micropropagation of pomegranate.

However, reports on pomegranate *in vitro* studies suggest the use of BAP or BA (0.5-5 mg/l) in combination with NAA (0.1-1.0 mg/l) for shoot bud differentiation and shoot regeneration in pomegranate while IBA and NAA (0.1 to 1.0 mg/l) are essential for good rooting. There are also reports on use of zeatin riboside (ZR), thidiazuron (TDZ) and Kinetin at concentration ranging from 0.5 -2.0 mg/l for shoot proliferation in pomegranate.

Other Additives: Activated charcoal (AC) is often added to plant tissue culture media because of its beneficial effects on many aspects of *in vitro* regeneration. AC effects have been attributed to various factors like darkening of the media, removal of inhibitory compounds, adsorption of agar impurities, adsorption of phenols produced by the tissues, *etc.* The positive effects of AC on the rooting of the microshoots can be attributed to reduction of light at the base of the shoots, thus providing an environment conducive to the accumulation of photosensitive auxin or cofactors besides, AC also adsorbs some of the components of medium and release them slowly thereby creating a stress like situation which helps in accumulation of root promoting factors in microshoots. PVP has also been included in media to avoid explant browning. Many a time ethylene inhibitor like $AgNO_3$ or AVG is used as media additive to inhibit ethylene activity as ethylene causes yellowing of leaves, inhibition of organogenesis, shoot growth, *etc.*

In vitro Hardening Strategies: Commercialization of *in vitro* propagation has been limited due to high field mortality and slow growth rate of tender plantlets during acclimatization phase. Underdeveloped and weak root system without proper vascular connection with the plant, unfavorable nutritional and environmental conditions, poorly developed leaf cuticle and absence of fully functional stomata are major hurdles in the successful field establishment of *in vitro* raised plants. *In vitro* rooted plantlets are transferred to sterilized potting mixture comprising of cocopeat + perlite + vermiculite (3:1:1) in glass jars either with poly-propylene cap or plastic pots with polythene cover. Before transfer, the basal portion of the plants including roots is washed with 0.1 % carbendazim to remove adhered medium and avoid fungal contamination. For one week they are kept in the growth chamber and then shifted to polyhouses with 16/8 light and dark cycle with 25±2ºC and 85-90 % relative humidity.

Biohardening: After 45 days of acclimatization, plant beneficial microbes (biohardening agents) are introduced in the root zone the plants. The soil and other carrier based Arbuscular Mycorrhizal Fungi (AMF) cultures (*Glomus intraradices/ Rhizophagus irregularis, Glomus mossae, Glomus manihotis, etc.*) and other plant beneficial microbes can be used for biohardening of *in vitro* raised plantlets under glasshouse conditions. Plant beneficial microbial formulations are placed in the root zone of plants at the time of their transfer to sterile potting mixture in nursery bags. These biohardening agents not only improve availability of nutrients to the plants by increasing root biomass but also help in improving immunity of plants system through induced systemic resistance.

Biohardening agents like AMF helps in improving the phosphorus nutrition and growth of the host plant, which may result in an increased resistance to various stresses. They increase root surface area for water and nutrients uptake. The use of AMF as biohardening agent helps in more branching of plant roots as the mycorrhizal hyphae grow from the root to soil enabling the plant roots to contact with wider area of soil surface, hence, increasing the absorbing area for water and nutrients of the plant root system. Therefore, plants with mycorrhizal association will have higher efficiency for nutrients absorption, such as nitrogen, phosphorus, potassium, calcium, magnesium, zinc and copper and also increased plant resistance to drought.

Besides, nutrient augmentation these microbes are responsible for anatomical changes in the root system, microbial changes in the rhizosphere and enhanced plant defense responses by altering the host's signaling pathways. Plant beneficial microbes like AMF plays critical role in increasing activity of hydrolytic enzymes, enhancing the levels of pathogenesis related proteins and accrual of phytoalexins in the plant system.

Effect of Biohardening Agents on Performance of Tissue Culture Raised Plants: An experiment on biohardening of *in vitro* raised pomegranate plants of cv. 'Bhagawa' had been laid out during 2012-13 at ICAR-NRCP. Two bio inoculants namely, *Arbuscular Mycorrhizal Fungi* (mixture of different types of AMF, predominantly *Glomus intraradices*) and *Aspergillus niger* (AN 27) were utilize as biohardening agents. Root colonization, population of microbes in rhizoshpheric soil, growth physiological and biochemical parameters of biohardened plants as influenced by these beneficial microbes were recorded at 180 days after inoculation. Colonization of roots of *in vitro* raised pomegranate plants with *Arbuscular Mycorrhizal Fungi.* (AMF) found at par in plants inoculation with AMF (71.12%) and plants inoculated with AMF+ *Aspergillus niger* (65.00%). Population of *Aspergillus niger* (AN 27) in the rhizospheric soil was found significantly higher in soil inoculated with *Aspergillus niger* (6×10^4 cfu/g of soil) as compared to non inoculated control (2×10^4 cfu/g of soil).

Plant height, shoot and dry weight and root fresh and dry weight recorded for establishment the superiority of biohardened tissue culture raised plants over non biohardened plants. Except for shoot dry height, which was found non significant among various treatments all other growth parameters were significantly influenced by biohardening agents (Table 10.1). AMF and AMD+Asp treated plants registered significantly better RWC (92.34 and 91.74%, respectively) and higher photosynthesis (12.69 and 12.78 μmol CO_2 $m^{-2}s^{-1}$, respectively) as compared to control and only Asp treated plants. However, performance of AMF inoculated plants either alone or in combination of *Aspergillus niger* excelled significantly as compared to control and only *Aspergillus niger* treated plants (Table 10.1).

Table 10.1: Effect of Biohardening on Growth, Physiological and Biochemical Attributes of Tissue Culture Raised Pomegranate Plant

Treatment	Plant Height (cm)	Shoot Fresh wt. (g)	Shoot dry wt. (g)	Root Fresh wt. (g)	Root dry wt. (g)	Leaf Relative Water Content (%)	Photo-synthesis (μmol $CO_2 m^{-2} s^{-1}$)	Total Phenol Content (mg Catechol equivalent / 100 g Fresh Wt.)
Control	110.69	157.09	61.13	49.84	14.63	87.76 (69.54)	9.07	29.50
AMF	140.63	196.03	74.25	61.44	16.92	92.34 (73.96)	12.69	52.50
Asp	130.38	183.23	73.14	60.40	16.04	89.99 (71.60)	9.70	42.00
AMF +Asp	134.63	184.23	74.94	58.44	16.30	91.74 (73.31)	12.78	53.00
CD(p=0.05)	15.25	23.81	NS	7.62	1.53	1.97	3.02	12.34

High photosynthetic activity of AMF inoculated and AMF + *Aspergillus niger* treated plants may be due to higher leaf chlorophyll content of plants under the influence of these two treatments as compared to remaining two treatments (Fig.10.1).

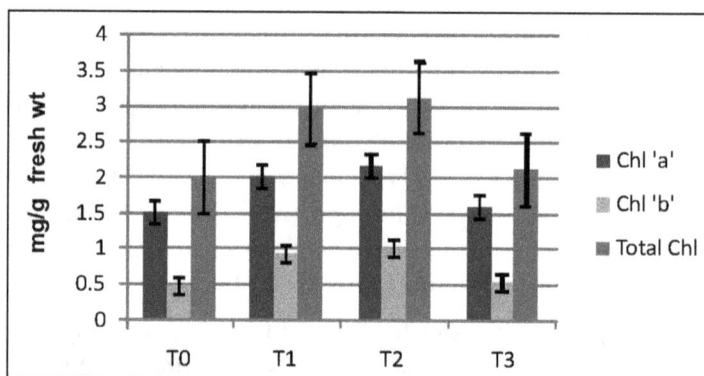

Fig. 10.1: Effect of Microbial Inoculation on Leaf Chlorophyll of Tissue Culture Raised Pomegranate Plants (T0: Control; T1: AMF; T2: AMF+Asp; T3: Asp)

Advantages of *in vitro* Propagation

- Availability of elite, disease free planting material in bulk
- Synchronized flowering and fruiting of *in vitro* raised pomegranate plants make them more suitable for mechanized cultivation
- The *in vitro* raised plants ensures high uniformity, precocity, better quality and yield as the mother plants of proven horticultural traits are used as the source explant

- Disease free elite planting material for extension of pomegranate to nonconventional areas as the method restricts the spread of pathogen through planting material

Conclusions and Future Research Needs

A judicious use of different propagation methods depending upon season, place, scale, initial capital and purpose has to be made in order to meet the increasing demand of quality planting material in pomegranate. Still, many aspects of effective propagation are unexploited in pomegranate at commercial scale which needs immediate attention. There is also an urgent need to develop nursery certification guidelines to avoid spread of insect-pests and diseases. Bio- hardening or bio-priming utilizing beneficial plant microbes are to be exploited for improving the field performance of *in vitro* raised plants. Even advanced protected propagation structures like mist chambers, shade net, polyhouse could be exploited for improving the propagation effeciency. Research on identification and utilization of potential root stocks must be intensified to cope up with various edaphic, biotic and environmental challenges in the regime of climate resilient horticulture.

11

Boron: A Critical Nutrient for Enhancing Pomegranate Productivity in India

M.V. Singh[1] and A.B. Singh[2]

[1]Agriculture Advisor, Rio Tinto and Former Project Coordinator (Micronutrients),
[2]Indian Institute of Soil Science (ICAR), Bhopal 462 043, India
Email: drmvsingh @ yahoo.com

Introduction

Micronutrients requirement of horticultural crops is very small but they play very important role in influencing crop growth, fruit set, yield and quality of produce. Pomegranate is commercially cultivated in Maharashtra, Andhra Pradesh, Rajasthan, Karnataka, Tamil Nadu, Gujarat, Odisha and Nagaland covering about 1.93 lakh ha producing about 22.0, lakh tonnes in 2015-16. Pomegranate is grown well under tropical and sub-tropical conditions. The soils of these regions have tested low to medium in fertility. Deficiencies of micronutrient like zinc, boron, iron, manganese are wide spread in above states (Singh 2011). Boron (B) plays very important role in number of metabolic pathways. It acts as fuel pump, aiding the transmission of sugars to new growth and root system. Its deficiency is increasing, so it invites much attention to study the impact of boron deficiency and assess management options influencing fruit quality and productivity of pomegranate in India. Important functions of boron in plants are as follows:

- Cell differentiation and development particularly the growing tips, xylems and phloem

- Increases flowering, elongation of pollen tube, pollination, fruit set, yield and quality

- Helps in sugar synthesis and translocation by forming sugar borate complexes in tips of plants

- As precursor of lignin synthesis through polymerization of phenolic compounds Helps in uptake of calcium and enhances NUE through balanced nutrient ratio. Enhances marketable fruit yield by decreasing physiological disorders to a great extent.

Boron deficiency leads to physiological disorders in plants. Terminal growth of twigs shows dies back, leads to partial defoliation and death of twigs. Leaf remains small, dwarf, thick and becomes brittle. New growth below dead tips either dies or forms a rosette. Blossoming is defective with poor pollination and fruit setting and plant bears small size fruits of poor quality. The cracking of fruits causes severe economic loss to the grower. Timely boron fertilization either in soil, or by foliar sprays or by fertigation are important practices to achieve best fruit production (Singh, 2011).

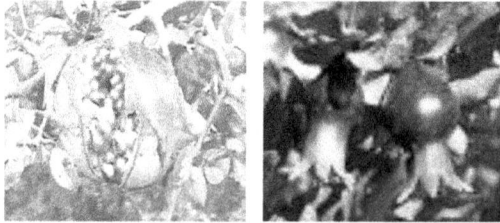

Plate 1-2: Effect of Boron Deficiency on Pomegranate Flower and Fruit Cracking

Boron concentration in plants depends upon the age and plant parts. Therefore, selection of right plant part at right growth stage is extremely important. For analysis of B in pomegranate, 8th leaf from apex at bud differentiation in April and August is found ideal. A concentration of 20-50 mg B/kg dry matter is found optimum but concentration below 20 mg B kg^{-1} in young leaf tissues is considered critical, below which growth of pomegranate is affected adversely (Tandon, 2009). Contrary to this, concentration of 200 mg B kg^{-1} reflects the boron toxicity above which reduction in potential growth of pomegranate tree is expected (Singh *et al.* 2006). Optimum calcium-boron ratio is reported to be between 200-280. Boron toxicity is rarely observed from excessive boron fertilization but its geogenic toxicity exists in several site specific areas of Rajasthan, Haryana, Uttar Pradesh, and Andhra Pradesh where orchards are irrigated regularly with saline ground water having high boron concentration (Mathur *et al.*,1964; Singh 2008, 2011).

Extent of Boron Deficiency in Soils

Boron deficiency in crops is widely observed because of adaptation of intensive planting of dwarf cultivars having higher B demand, extensive plantation on marginal lands, low use of manures, and other soil-crop-management factors that limit adequate boron availability to plant and creates nutrient imbalances (Singh, 2001). Fluctuating soil moisture, quality of irrigation water and geogenic factors also affect forms and distribution of B in soil (Singh and Saha, 1995). Boron deficiency in soils is reported when hot water soluble boron (HWS B) level is below 0.50 mg kg^{-1} soil whereas boron toxicity was reported when HWS-B concentration is above 3 mg kg^{-1} soil (Bell 1997, Singh 2009). Available (HSW) B) status in Indian soils ranged from 0.05-8.2 mg kg^{-1} with a mean of 0.77 mg kg^{-1} soil though total boron ranged between 7 to 630 mg kg^{-1} (Singh, 2008). Only about 10% of total B is likely to be available to the growing plants depending upon soil-crop-management factors.

Boron deficiency is not only impacting adversely on pomegranate production in India but its deficiency is reported in sixty countries of the globe (Bell 1997, Shorrocks 1997, Singh 2006, 2009). Deficiencies of B and Zn could be suspected in almost every country. Singh and Goswami (2014) estimated that 52% soils of India are deficient in B though initially in eighties it was reported only in 2% soils (Kanwar, 1976). Average boron deficiency has spread as much as in 60.8% in soils of southern states, 46.8% in eastern states, 30.68% in western states and the least 13.86% in soils of northern India as shown in Fig. 11.1, 11.2 (Singh 2011).Thus, B deficiency has become one of the serious constraints of similar magnitude to that of zinc in getting higher horticultural crop production in the country.

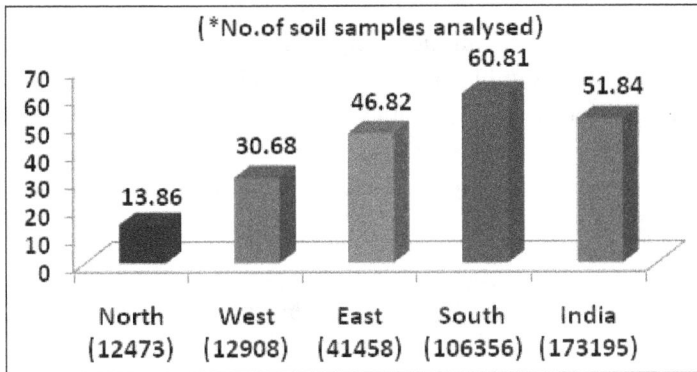

Fig. 11.1: Percent Boron Deficiency in Soil of India

State wise boron deficiency is found in more than 60% soils of Andhra Pradesh, Karnataka, West Bengal, 45-55% in soils of Odisha and Jharkhand, and less than 35% in soils of Madhya Pradesh, Maharashtra, Rajasthan, Tamil Nadu and the least 8-15%

deficiencies in soils of Haryana, Punjab and Uttarakhand (Singh and Wanjari, 2013). It also affects the crop production in 38% calcareous soils of Bihar and 45% acid soils of Jharkhand. The deficiency of B is reported extensively in soils of Karnataka, Maharashtra, Andhra Pradesh and Madhya Pradesh. Horticultural crops are more vulnerable to its deficiency. So B deficiency is recognized an important nutritional constraint in many crops, soils and states of India (Wani *et al.* 2011, Singh 2006).

Boron toxicity is not a common problem in India so the focus always remained on management of deficiency. Boron toxicity is reported in areas of salt affected soils or areas having high B saline ground waters. Geogenic B toxicity caused by "inherent site specific soil- water factors" invites much attention to manage as its effect on crop growth is far severe due to constant prolonged supply of boron from irrigation water to the root zone and soluble boron build goes above 3 mg/kg soil.

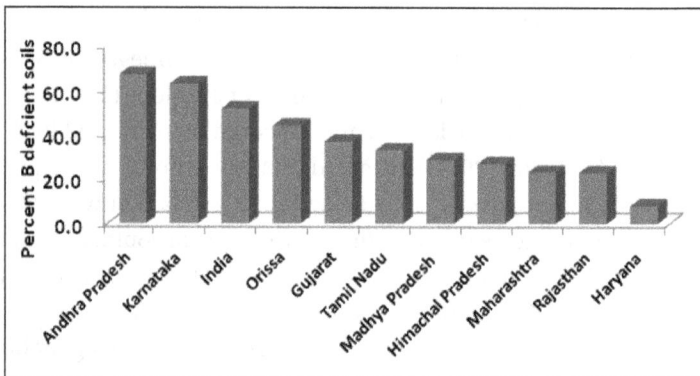

Fig. 11.2: Percent Boron Dificiency in Various States

Excess accumulation of B induces phyto toxicity as leaf necrosis; reduces leaf area to the extent which significantly reduces the photosynthesis, yield and quality of produce in common crops. Preventive approaches to reduce boron toxicity like growing B tolerant crops, cultivars or less irrigation/ water demanding crops, organic manuring, use of amendment, higher use of N P K Zn and leaching are preferred.

Management of Boron

Although all horticultural crops respond significantly to micronutrient fertilization but management of boron deficiency is critical in almost all orchards which affects growth, flowering, fruit set, yield and quality of fruits. All boron sources as approved in Fertilizer Control order (FCO) are given in table 11.1.

Table 11.1: Sources of Boron Approved in FCO, their Composition and Rate of Application in Crops

Chemical name	Fertilizer Chemical form	Chemical Formula	B content percent as per FCO	Fertilizer doses, kgha^{-1}	Application Mode
Disodium tetraborate penta hydrate	Granubor° (Granular)	$Na_2B_4O_7.5H_2O$	14.6	3-7	Soil, basal /ferti. blending
Disodium tetra borate penta hydrate	Fertibor°* (Crystalline)	$Na_2B_4O_7.5H_2O$	14.6	3-7	Soil / foliar spray/ fortification
Disodium tetra borate deca hydrate	Borax (Fine crystal)	$Na_2B_4O_7.10H_2O$	10.5	7-10	Soil
Disodium octa borate tetra hydrate	Solubor° (Powder)	$Na_2B_8O_{13}.4H_2O$	20.0	1-2 g/L of water	Foliar spray
Boric acid fine	Boric acid (Crystal)	H_3BO_3	17.0	1-2g/L of water	Foliar spray

In pomegranate, application of 1-2 g borax penta hydrate (15% B) per square meter canopy as basal and foliar spray of 1 g borax octa hydrate per litre solution two to three times increased the plant growth, bearing and size of fruits. However, foliar sprays of combined application at 0.4% ferrous sulphate, 0.3% zinc sulphate, 0.3% manganese sulphate and 0.1% disodium octaborate tetra hydrate solution neutralized with lime twice enhanced the fruit yield significantly over individual sprays of Zn, Fe, Mn and B alone (Table 11.2).All boron source viz. granubor, fertibor and borax can correct boron deficiencies in soils (Singh 2006, Singh and Goswami 2013). Calcium borate (Colemanite) is however, less efficient source of B due to least water solubility in calcareous, alkaline alluvial and swell shrink soils. New generation products like disodium octaborate tetra hydrate (Solubor°, 20% B) and boric acid (17% B) were found better soluble boron source for foliar sprays due to their instant solubility in water. Foliar sprays gave higher yields, reduced fruit cracking, increased size and shining of fruits as well as quality of juice. Increasing concentrations of trace elements significantly increased the sugar content, TSS and reduced the acidity in pomegranate. Borax penta hydrate is easy to apply directly to all crops/soils or through boron fortified fertilizers. Government has also approved boron fortification in the range of 0.1-0.3% depending upon nature of P/NP/NPK fertilizers. Promoting boron fortified fertilizers would be the successful way forward for correcting boron deficiency and raising fruit yields of pomegranate.

Table 11.2: Response to Foliar Spray of Micronutrient in Swell-Shrink Soil of MP

Treatment*	No.of fruits Per plant	Fruit yield (kg/plant)	Mean volume of fruit (cm^3)	Juice Percent	Increase in yield over control (%)
0	65.2	21.0	262.3	65.3	-
B	75.2	28.2	331.5	71.4	34.29
B+Fe	81.2	30.2	348.7	74.3	32.62
B+Fe+Zn	88.7	33.6	360.9	75.6	41.72
B+Fe+Zn+Mn	93.6	35.6	365.3	78.7	43.45
LSD 0.5%	12.8	4.6	27.4	2.3	21.9

*Two foliar sprays of micronutrient mixture *viz.* 0.1% disodium octaborate tetra hydrate, 0.4% ferrous sulphate, 0.3% zinc sulphate, 0.3% manganese sulphate solution neutralized with 0.5% lime

Effect of Organic Vs Inorganic Fertilization on Pomegranate

Field studies conducted in swell-shrink soils of Bhopal to compare efficiency of organic, inorganic and integrated nutrient management systems on the yield and quality of pomegranate revealed, that organically treated trees gave less number of fruits but of bigger size, had higher sugar and low acidity in juice whereas number of fruits per tree were higher in case of chemical fertilizer treated trees but fruits remained smaller in size and infected. Application of 18 kg vermicompost or 21 kg compost or 29 kg cattle dung manure (CDM) per tree or a combination of 50% CDM+50% RDF improved the average fruit weight and fruit yield (Table 11.3). Over all trends of pomegranate yield was found in the order of: Integrated nutrient management > organic nutrient sources > chemical fertilizers > no fertilizer. Organic manuring also increased the yield, size of fruit, juice, ascorbic acid, pH and TSS. Regular use of 10-15 t ha^{-1} manure, green manure, vermicompost can prevent emerging boron deficiency in various soils.

In another study, use of well decomposed 10 kg FYM/tree + recommended dose of nutrients annually to soil and foliar spray of multi micronutrients (including 2 g/L solubor) on pomegranate at full blossom stage gave best yield, increased ascorbic acid, pH and TSS. Efficiency of nutrients applied only to soil were found less efficient than fertilization by drip whereas drip fertilization plus foliar spray of micronutrients gave best fruit size, shining and maximum fruit yield in black clayey soils. Micronutrient boron is compatible with other soluble fertilizers. However, compatibility of fertilizer solution should be judged with irrigation water quality. Mixing it with oily substances is not compatible. Drip irrigation- fertilizer system needs to be flushed with water to avoid clogging.

Table 11.3 : Effect of Different Manure on the Yield and Yield and Quality of Fruits of Pomegranate in Swell-Shrink Soil of Bhopal

Treatments	Plant height (m)	Plant spread (m)	No. of fruits/ plant	No. of infected fruits/ plant	Average Fruit wt (g)	Fruit yield kg /plant	Juice recovery (%)	Sugar (%)
Control	3.1	4.7	31	8	150.5	4.7	41.0	11.2
Vermi - compost	3.3	5.3	40	9	177.8	7.1	55.8	12.7
Phospho- compost	3.4	5.5	40	7	184.1	7.3	55.4	12.6
Cattle dung manure	3.3	5.7	39	7	185.6	7.2	59.8	13.0
RDF inorganic	3.7	5.2	44	13	155.4	6.8	45.9	11.7
50% RDF+50% CDM	3.6	5.7	42	8	185. 6	7.8	58.3	12.9
CD (P=0.05)	NS	1.10	2.3	--	6.70	1.8	3.26	0.09

Summary: Meager information is available on the prevalence of boron deficiency in pomegranate orchards and suitable management options despite good visible boron deficiency effect in the form of physiological disorders on plant growth and poor quality of fruit as cracking is seen widely. Paper summarizes effect of boron on fruit set, yield and quality of pomegranate.

Considering higher requirement boron fertilization is essential to all pomegranate orchards. Soil application and foliar sprays of boron were found much beneficial. Combined application of multi micronutrients (Fe, Zn, Mn, B) gave maximum yield and quality fruit than spray of either of nutrients. New generation boron products like granubor is more efficient than other sources due to slow release properties in soil whereas solubor powder is best for foliar sprays due instant solubility in water or by drip fertilization. So, boron fertilization is crucial for achieving higher economic returns and enhancing yield of pomegranate by 20-35% over no boron treatment. Under organic farming system pomegranate gave lower yields but of bigger size less infected fruit whereas reverse trend was found for inorganic fertilized tree. Boron toxicity due fertilization is merely observed where as sporadic geogenic toxicity due to high boron containing saline waters in the states of Andhra Pradesh, Gujarat, Uttar Pradesh, Haryana, Punjab is noticed. Therefore, boron fertilization should be practiced by considering salinity level of ground water. Fertigation / drip irrigation are found good for applying small amount of boron as per critical stages of plants.

12

Effect of Elemental Sulphur on Solubility of Micronutrients and their Uptake by Pomegranate

Ashis Maity, Ram Chandra, N.V. Singh,
D.T. Meshram and R.K. Pal

ICAR-National Research Centre on Pomegranate, Solapur-413 255, Maharashtra
E-mail : ashisashismaity@gmail.com

Introduction

Pomegranate (*Punica granatum* L.) is one of the economically important fruit crops of the world widely under tropical and sub-tropical region (Jalikop, 2007).This crop is a good source of protein, carbohydrate, minerals, vitamins A, B and C and is well known for its anti- cancerous, anti-microbial, anti-oxidant and anti-inflammatory properties (Albrecht *et al.*, 2004; Lansky and Newman, 2007; Opara *et al.*, 2009). Because of its drought tolerance and winter hardiness characteristics it can thrive well under desert condition. In India it is commercially grown in semi-arid regions (Deccan Plateau) which is characterized by high pH, high exchangeable sodium, varying degree of calcareousness, excessively low permeability and poor fertility. Because of high pH and free $CaCO_3$ content of soil the availability of micronutrients to the plant get adversely affected and plant suffer from micronutrient deficiency unless they are supplemented trough foliar application. The micronutrients *viz.* Fe and Mn which are phloem immobile, it is very difficult to enhance their content in fruit through foliar application of those micronutrients. They are mainly transported within the plant via xylem in transpiration pull. For enhancing the content of Fe and Mn in the fruit, it is very much imperative to solubilize soil reserve of those micronutrients through localized reduction of rhizosphere soil pH. The reduction of soil pH could be accomplished through replacing sodium from soil exchange sites to a level safe for crop growth. Gypsum is commonly used for reclamation of alkali soils owing to its easy availability and low cost as compared to other chemical amendments. As the pomegranate growing soils are rich in free $CaCO_3$ content, use of elemental sulphur (S) to decrease soil pH and increase the solubility of micronutrients in soils would be more effective than gypsum (Tichy *et al.*, 1997; Kayser *et al.*, 2000). Besides, the elemental S will full fill the nutritional requirement of the plant as this nutrient is generally neglected in pomegranate production. The present study was planned to investigate the effect of elemental S on plant growth, physiological activity, availability of soil micronutrients and their uptake by the plant in non-calcareous and calcareous alkali soils.

Materials and Methods

To study the effect of elemental S on plant growth, micronutrient availability and their uptake by the plant a pot experiment was conducted for 15 months at ICAR-NRC on pomegranate experimental farm, Solapur on two types of alkali soils viz. non-calcareous soil (initial pH 7.68) and calcareous soil (initial pH 8.05) collected from the top 60 cm of the pomegranate orchard soil. The experimental designs were employed with elemental sulphur and zinc (Zn). Sulphur was applied and mixed thoroughly with the soil at four rates, 0 (S_0), 2.5 (S_1), 5.0 (S_2) and 10.0 (S_3) g kg^{-1} soil and Zn applied with two rates, 0 (Zn) and 25.0 (Zn) mg kg^{-1} respectively. There were 16 treatments in total and each one with three replicates. The Zn was thoroughly mixed with soil after elemental S was added. Four months old uniform saplings of pomegranate cv. Bhagawa were planted in soil in pots of 8 kg capacity. Recommended dose of fertilizers was applied. The irrigations during the study period were applied on need-felt basis.

Measurements on chlorophyll content (using chlorophyll meter, KONICAMINOLTA SPAD-502 as indicated by SPAD value) and photosynthetic rate (using LI-COR LI-6400 portable photosynthesis system) were recorded at 12 months after planting (Field *et al.*, 3).

Fifteen months after treatment, the whole plant was up-rooted and plant biomass was dried at 105°C after separating the shoot, root and soil. The criteria for growth promotion were studied as root and shoot dry matter. Plant micronutrient (Fe, Mn,

Zn and Cu) content was measured after acid digestion according to methods of the Association of Official Analytical Chemists (2). Soil samples were taken from the middle of each pot at harvesting time to assess the treatment effect on soil properties and available micronutrient content. The soil pH, electrical conductivity, DTPA extractable micronutrients were determined by standard procedures (Jackson, 1967).

Results and Discussion

The study indicated that application of elemental S significantly reduced pH from 7.68 to 5.14 in non-calcareous loamy soil and from 8.05 to 7.58 in calcareous clayey soil (Fig. 12.1). The degree of reduction of soil pH was higher in non-calcareous loamy soil. One important fact is the oxidation of S by certain groups of acidophilic bacteria, notably *Thiobacillus* spp. in the soil (Lee *et al.*, 1988; Kayser *et al.*, 2000). Kayser *et al.* (2000) reported that adding 36 mol S m-2 to the soil led to a decrease in soil pH from 7.2 to 6.9. In the present study, application of S also acidified the soil which caused the soil pH to decrease about 2.54 and 0.47 unit in non-calcareous clay loam and clay soil respectively. Conversely, significant increase in electrical conductivity was observed in calcareous clayey soil owing to its poor drainage characteristics (Fig. 12.1).

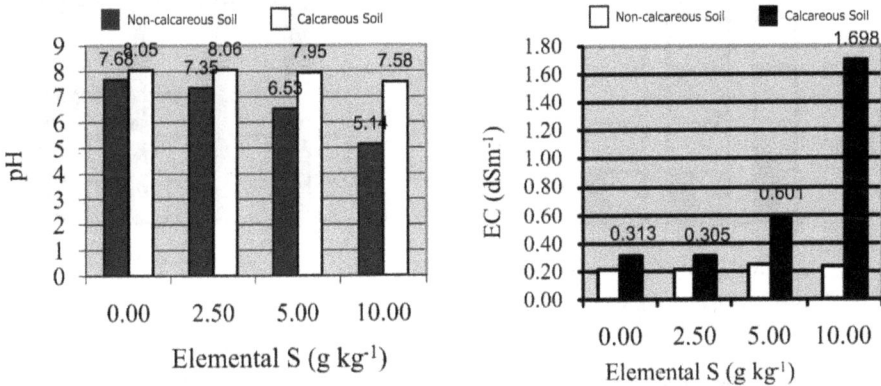

Fig. 12.1: Effect of Elemental S on Soil pH and Electrical Conductivity

Lowering of soil pH upon application of elemental S significantly increased available Fe and Mn status in both the soil, however the extent of increase in availability of Fe and Mn was higher in non-calcareous loamy soil than calcareous clayey soil (Fig. 12.2). Application of elemental S also enhanced available Zn status of calcareous clayey soil. The mobile fraction of soil micronutrients is greatly influenced by soil pH and generally increases as soil pH decreases (Kayser *et al.*, 2000; Martinez and Motto, 2000) Although, application of Zn @ 25 mg kg^{-1} soil significantly increased available Zn status in both the soils but significant negative interaction between elemental S and Zn was observed in both the soil (Fig. 12.3).

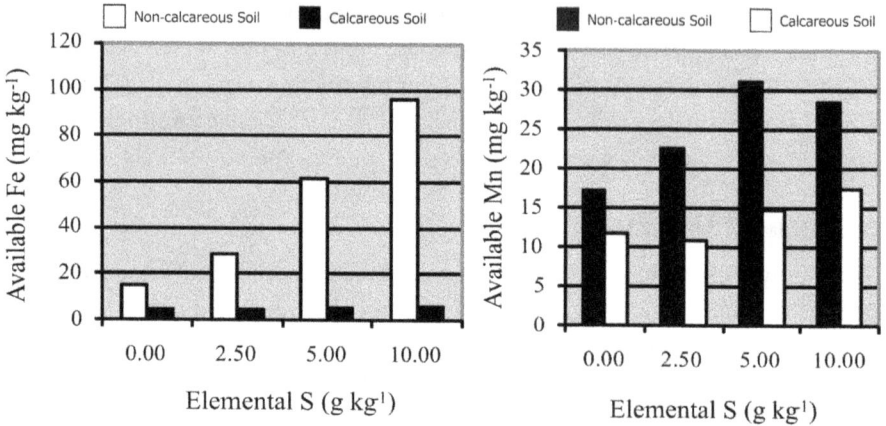

Fig. 12.2: Effect of Elemental S on Available Fe and Mn Content of Soil

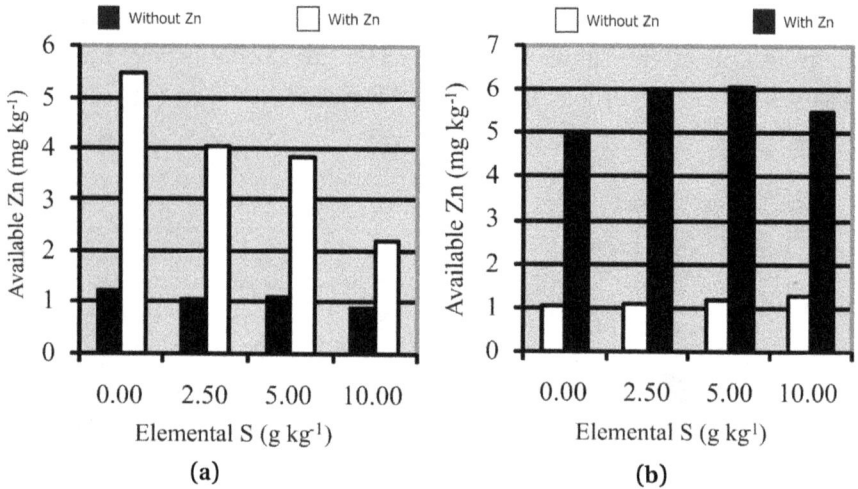

Fig. 12.3: Effect of Elemental S and Zn on Available Zn Content of (a) Non-Calcareous Loamy Soil and (b) Calcareous Clayey Soil

There was significant increase in Fe and Mn uptake by the plant with the application of elemental S in non-calcareous loamy soil and the maximum increase was observed when elemental S was applied @ 5.0 g kg^{-1} soil (Fig. 12.4 & 12.5). In calcareous clayey soil application of elemental S also increased Fe and Mn uptake by the plant however, maximum uptake was observed with elemental S when applied @ 10.0 g kg^{-1} soil. Further, application of Zn @ 25 mg kg^{-1} soil significantly increased Fe and Mn uptake by the plant but this extent of increase reduced with the increasing rate of elemental S application. Here, negative interaction was noticed.

Application of elemental S significantly increased Zn uptake by the plant in both the soil however, maximum increase in Zn uptake was observed with 2.5 g S kg^{-1} soil in non-calcareous loamy soil whereas in calcareous clayey soil it was with

10.0 g S kg^{-1} soil (Fig. 12.6). There are number of reasons for this result in high plant uptake and accumulation of these micronutrients; however the main reason is most probably that the increase of solubility of Fe, Mn and Zn leads to an improvement in metal bioavailability with a decrease of soil pH by application of S. There are certain groups of acidophilic soil bacteria in the soil, predominantly the genus *Thiobacillus*, that can oxidize S and change it to SO^{2-} when the soil pH has decreased (Lee *et al.*, 1988; Tichy *et al.*, 1997; Kayser *et al.*, 2000). Further, application of Zn @ 25 mg kg^{-1} soil significantly increased Zn uptake by the plant in both the soils however, positive interaction was observed in non-calcareous loamy soil and negative interaction was observed in calcareous clayey soil.

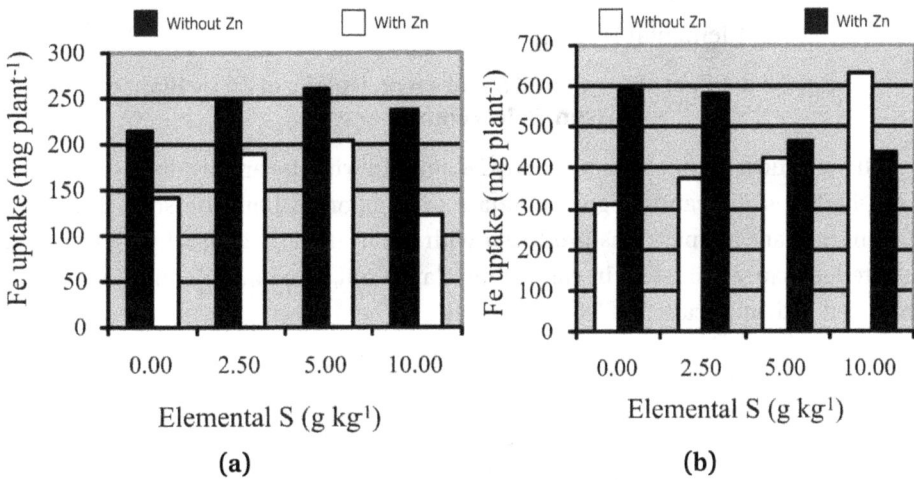

Figure 12.4: Effect of Elemental S and Zn on Uptake of Fe by Plant on
(a) Non-Calcareous Loamy Soil and (b) Calcareous Clayey Soil

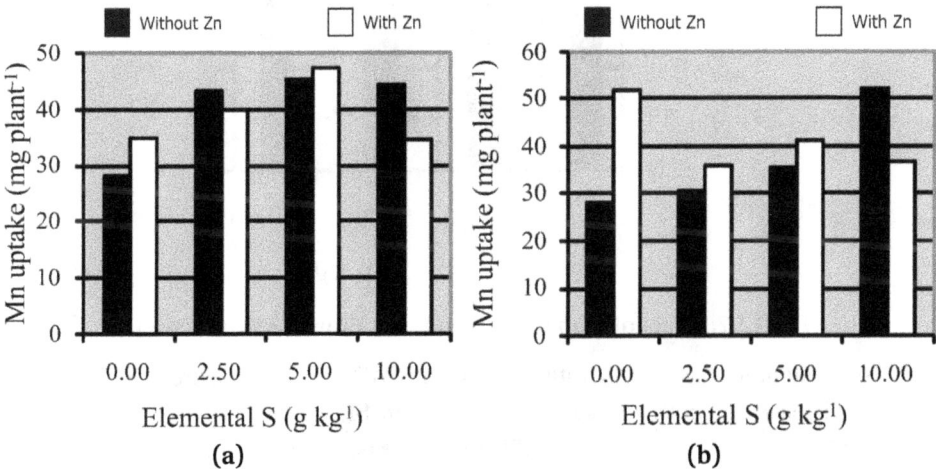

Figure 12.5: Effect of Elemental S and Zn on Uptake of Mn by Plant on
(a) Non-Calcareous Loamy Soil and (b) Calcareous Clayey Soil

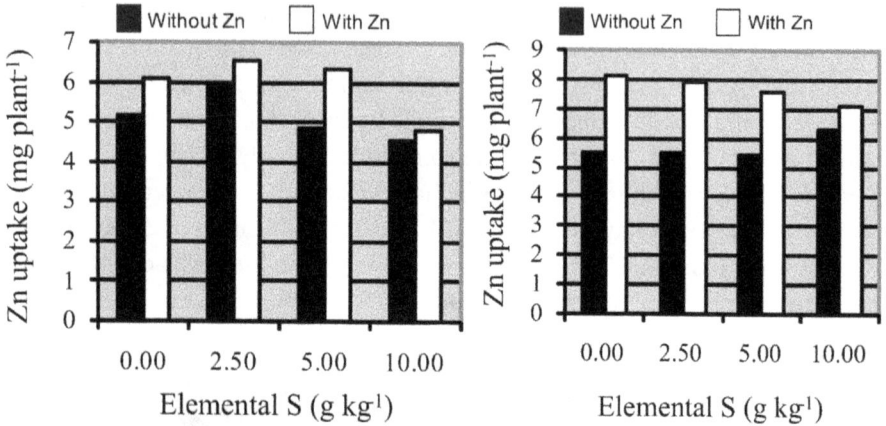

Fig. 12.6: Effect of Elemental S and Zn on Uptake of Zn by Plant on (a) Non-Calcareous Loamy Soil

Higher nutrient uptake particularly Fe and Mn with the application of elemental S resulted in significantly higher biomass production in both the soils (Fig. 12.7). Maximum plant biomass was produced with application of 2.5 g S kg^{-1} soil in non-calcareous loamy soil while in calcareous clayey soil; maximum plant biomass was produced with application of 5.0 g S kg^{-1} soil.

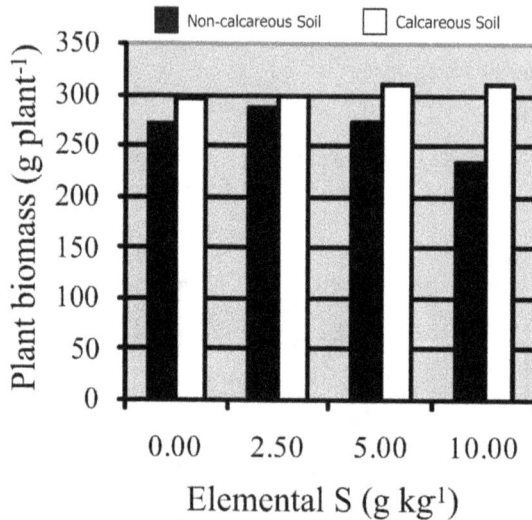

Fig. 12.7: Effect of Elemental S on Plant Biomass Production

Use of elemental S significantly increased SPAD value (measure of chlorophyll content of leave) in both the soil and maximum SPAD value was noted with the application of 5.0 g S kg-1 soil (Fig. 12.8). This increase in SPAD value in-turn resulted significant increase in photosynthesis rate in non-calcareous loamy soil and maximum photosynthesis rate was noted with 2.5 g S kg-1 (Fig. 12.9).

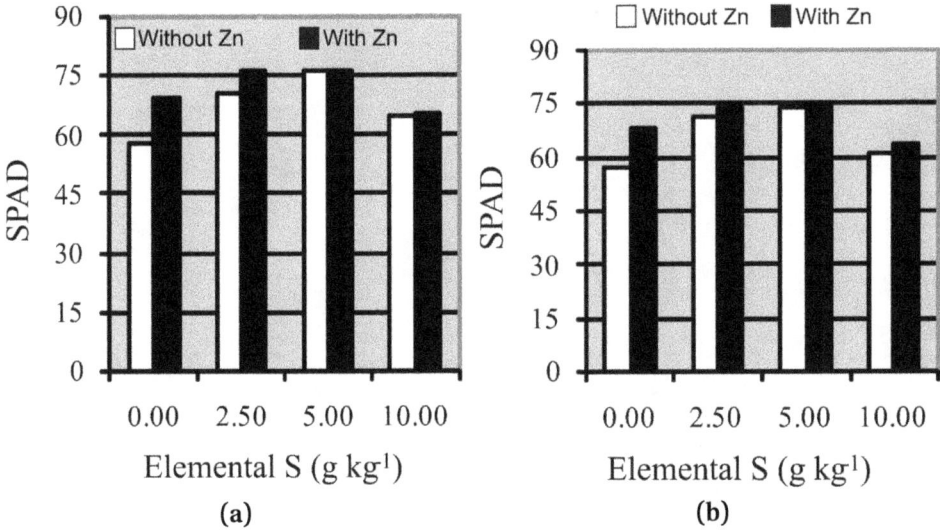

Fig. 12.8: Effect of Elemental S and Zn on Chlorophyll Content as Measured
by SPAD Value of Plant on (a) Non-Calcareous Loamy Soil and
(b) Calcareous Clayey Soil

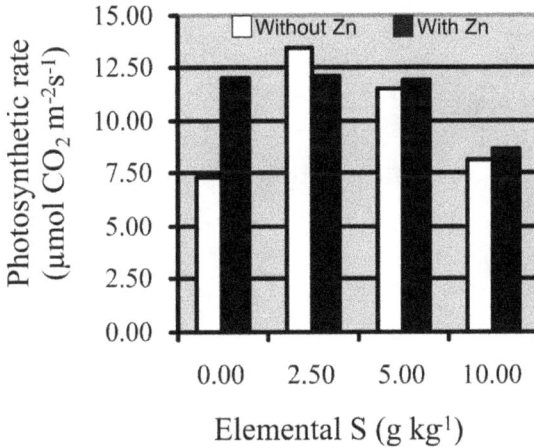

Fig. 12.9: Effect of Elemental S and Zn on Photosynthetic Rate of
Plant on Non-Calcareous Loamy Soil

Application of elemental S@ 5.0 g kg-' soil in non-calcareous loamy soil and@ 10.0 g kg-' soil in calcareous clayey soil resulted maximum increased in Fe and Mnuptake by tbe plant. Soil application of Zn @ 25 mg kg-' soil also significantly increased Zn uptake by tbe plant in botb tbe soil types, however negative interaction with elemental S was observed.As a result of higher nutrient uptake by tbe plant there was significant increase in plant growth as indicated by biomass production, chlorophyll content of leaves as indicated by SPAD value and photosynthetic rate of plant.

13

Rejuvenating Old Senile Pomegranate (*Punica granatum* L.) Orchards for Enhanced Quality Production

Rajesh Kumar

ICAR-National Research Centre on Litchi,Muzaffarpur-842 002, Bihar
E-mail: rajeshkr_5@yahoo.com

Introduction

Pomegranate (*Punica granatum* L.) commonly known as '*Anar*', is an ancient favourite table fruit of tropical and subtropical regions of the world. It belongs to family – Lytheraceae. There is a great diversity in perennial fruit crops in our country, among which the pomegranate is commonly finding place. Fruit orchard of this important fruit crop is somewhat expensive and skillful job but high productivity of its fruit trees gives higher economic return per unit area apart from being the means of livelihood security and usefulness in the preservation of the environment, like other commercial fruit crops. It is logical that higher production also increases the production of higher nutraceuticalhealthy constituents for human consumption from the unit area. As the red-juicy multi seeded fruits of the pomegranate are not only beautiful, but they are packed with nutrients. Pomegranate takes longer period to come into bearing stage (3-4 years) and thereafter due to neglect in management practices the annual production,the yieldgradually declines in degraded lands but, the successful commercial fruit production andincreased quality production has made significant contribution to economic development of the grower and ultimately those in the growing region in sustainable manner.

Need of Rejuvenation

In case of pomegranate too, the trees witness decline in productivity after certain age making orchard economically non-viable and non-remunerative. In declining orchard, the productivity status is further affected due to compounding problem of insect-pests and diseases. Those trees/plant are rejuvenated which have attained a stage where they are no more profitable from the grower's point of view (Singh, 2007).

It is required to take up productivity improvement programmes in the senile plantations, with fresh stock supported with appropriate and integrated combination of inputs, pruning and training techniques. This requires proven technology implementation through individual farmers, farmers' cooperatives, self-help groups, NGOs, growers' associations and commodity organizations. The technology for transformation of old senile orchards for enhanced quality fruit production requires well planned and concerted efforts. Though processes and procedures require scientific skill with precision in application/implementation.

The research efforts initiated at various pioneer institutes/ SAUs to standardize a technology for restoring the production potential of existing plantations by a technique called rejuvenation. Rejuvenation of old orchard is very vital under orchard management, which needs careful, visionary and scientific approach and it should be followed in skillful manner. The hard core pruning and proper training is a medium term strategy requires scientific skill and technique (Hiwale, 2009). The existing orchards turned or turning to unproductive stage and uneconomical can be brought back to productive ones through rejuvenation technique, which in turn help in restricting the deterioration of the health of the plant, thereby extending their life and aims at improving the yielding capacity of quality production of the existing old plants/trees having renewable ability, under minimum period of time as compared to establishment of new orchard.

Rejuvenation Technique

For the successful translation of the rejuvenation technology in old senile orchards the sequential operational steps should be followed in scientific manner, which include:

1. Identification/Selection of Old Orchards

The old orchards become non-productive due to incidence of pest and diseases and neglect are selected. Once the orchard becomes old its productivity goes down, and it is general tendency of the farmers to neglect the orchard. In pomegranate when the survey was conducted it was observed that 25 per cent of the old orchards become unproductive due to loss of vigour in the plantsand neededadoption forrejuvenation technology. The cause of decline should be determined and corrected if possible; otherwise, response to rejuvenation pruning will be temporary. If the trunk and basic scaffold limbs of trees are not structurally sound due to disease, heart rot, and cankers, such rejuvenation procedures are not justified. A limited root system due to a hardpan or poor subsoil can result in early tree decline, with trees being more readily stressed by drought. When this occurs, the top becomes out of balance with the more limited root system with consequent insufficient uptake of water and nutrients by the roots. Hence, selection of old orchards for rejuvenation is also a visionary approach.

2. Training to the Orchard Owner/Awareness to the Grower

Creating awareness for increasing adoption of this technology at the initial level needs proper training. As in pomegranate, rejuvenation can be done by mechanical measures, chemical measures as well as by hard core pruning of plants (Hiwale, 2009). Rejuvenation technology (an invasive technology) for old senile fruit trees, particularly in pomegranate, done through hard core pruning presents many (a variety) cross cutting issues on the theme of restoration of youthful vigour, enhancement of quantum of quality fruit production, period of sustainability and above all the economic feasibility, which encompasses aspects of scientific, suitability, productivity, sustainability and profitability. All these require proper implementation leading to success stories. This concern for the awareness to the growers and enough knowledge base through training, convincing the economic feasibility is very vital to increase the level of adoption.

3. Hard Core Pruning and Training Operations

Pruning plays an important role in regulating and controlling growth, flowering and fruiting in perennial fruit trees (Crane *et al*, 2009, Singh *et al*, 2012). Hard core pruning for topping off these trees should temporarily restore them to a more favourable top to root ratio. The severity of pruning for rejuvenation will depend on the cause and degree of decline. The cutting back the scaffold limbs to a height of 0.30 to 1.00 m above the ground, is the most severe form of rejuvenation pruning for this crop. It greatly reduces tree size and results in vigorous re-growth. Large wounds remain and production is lost for about one year. It is advisable to paint large cut surfaces on or near the trunk of the tree which result from hedging and topping. Severe topping for rejuvenation which exposes large limbs and trunks that have grown in the shade may result in severe sunburn when done during or shortly before hot, dry weather. In principle, prune perennial fruit trees when they are in dormant phase. In case of rejuvenation, it is important to look for the health, location and value of the fruit trees, which should be determined before going for rejuvenation. Work carried out on non-selective pruning at CHES, Vejalpur has shown encouraging results in case of hard core pruning in fifteen year old orchard that was showing signs of decline mainly due to drying of old branches resulting in sparse flowering and fruiting. The plants were headed back to 30cm, 60cm, 90cm and 120cm above ground level in the month of April, no pruning acted as a control. Training adopted as new sprouts per plant was reduced to 8 to 10, by allowing two to three healthy sprouts per branch to balance the framework of the plant. The treatment plants pruned to 30 cm from the ground level showed the best performance. Yield per plant was however surpassed control in second year itself. Pruning plants to 30 cm from ground level was the best to rejuvenate the old pomegranate orchard and can be recommended as standard practice to rejuvenation old orchards (Hiwale, 2009).

Many mistakes are made when people prune large or heavy branches. In many cases, the pruning process often results in damage to the tree. Pruning of limbs should be careful in order to avoid tearing or ripping of the bark while making the cut. When single cut in a hurry is used, the problem occurs. The final and finishing cut is made at the trunk marked at the desired height, since the weight of the limb/branches has been removed, this final cut can be made with precision and without the risk of damage to the bark of the trunk. Pruning can be done either with manual saw or power operated saw preferably in phase manner from the top. Use of pruning paints and dressings at the cut/wound portion is very important to act as defenses against disease and pest infestation. The understanding about the basics of pomegranate trees, when and how to prune is must. The process must be started soon after the emergence of new sprouts, as in this tree, there is a strong desire for a tree-form. The bushy, free-growing shrub develops naturally leading to many unproductive branches. Pomegranate fruits are borne terminally on shoot spurs, arising from matured shoots which have the capacity to bear fruits for 1-2 years. With advance in age, they decline in productivity. Later, a little thinning and pruning of old spurs to encourage growth of new ones is required. Farmers should always have the idea of some useful tips on pomegranate pruning.

4. Nutrient and Water Management

Nutrient and water management in case of rejuvenated trees are very vital, as it help in restoration of proper growth and vigour for sustainable fruit production in a long way. Pomegranate productivity is greatly increased by application of manures and fertilizers. Both macro and micro-nutrients affect its growth, development and productivity. After top off pruning for rejuvenation, the half of the recommended adult basal doses of farmyard manure and NPK should be applied after a month gap only. These fertilizers are applied in shallow circular trench of 8-10cm depth around the main tree trunk of half meter radius. Application of about 20kg FYM each tree every year at the onset of monsoon has been found beneficial (Gaur *et al.*, 1971). After application, the fertilizers are covered with top soil and the plants are irrigated for getting good invigoration effect and early emergence of sprouts. The soil should be well drained. Water management varies from season to season. Ordinary irrigation practices are quite enough for pomegranate fruit production but drip irrigation be followed to economize water. At present, it has become the major irrigation system for pomegranate fruit cultivation in Maharashtra.

5. Intercropping

During initial 2-3 years after rejuvenation, there is excellent opportunity for economic utilization of inter-space in the rejuvenated orchards as just after the completion of reiterative pruning for rejuvenation, the open interspaces between the plants/trees is created in the orchard like newly planted orchard. Intercrops during summer season and during rabi season have been found most suitable. Apart from the significant income obtained by the intercrops in rejuvenated orchards, the added advantages like improvement in the soil tilth (condition), almost complete check over weed population/growth as well as fewer incidences of pests – diseases are also achieved. Water requirements of the intercrops should not clash with those of the main fruit trees. The intercrops selected should not exhaust the soil water and nutrients and should not demand more water than is allowed for rejuvenated fruit trees. The most important aspect is that intercropping should be started atleast after one month, when hard core pruning operation is over. The rejuvenated pomegranate trees start commercial bearing after 2 to 3 years and until then intercropping can be done with fodders like berseem, lucern, leguminous crops like cowpea, green gram and vegetables like cabbage, cauliflower, beans, peas, tomato, carrot, onion, radish, potato, brinjal, *etc.* However, intercrops should be regulated so that their cultural care does not interfere with the bahar regulation at start of bearing stage. Similarly the intercrop should not help in pest carryover leading to heavy pest incidence on the main crop.

6. Mulching

To avoid occurrence of physiological disorders at after attaining fruit bearing stage, the mulching operation during the summer months is essential, apart from the nutrient and irrigation management, which is mainly to avoid the variation in moisture content of the soil as well. The mulching also takescare of the thermal regulation of the rhizosphere region and weed population control (Sadhu and Gill, 2013).The standard plant protection measures should be followed.

Fruit Yield

Fruit yield and physicochemical characteristics of mature fruits were found to be better in fruits obtained from rejuvenated trees. Maturity period was found to be slightly delayed in rejuvenated plants. The tree starts yielding from second year onwards giving 70-75 fruits per tree. Yield increases progressively from 3rd to 5th years and a tree may produce on an average 175-200 fruits per year. The average yield in a further well managed orchard may be as much as 200-250 fruits per tree. Economic yield obtained continuously for another 10-15 years after rejuvenation.

Economics of Rejuvenation Technology and Impact

The various operations in rejuvenating old senile trees is not an immediate process, but usually it can be done effectively over a period of 2-3 years and including maintenance and management aspects. The cost of hard core pruning and subsequent operations in one hectare of old senile orchard have been estimated (Table 13.1). It has been found that the old senile orchard can be rejuvenated back just like young bearing orchard within a time gap of 2-3 years without any income loss in-between by adopting and following the technology. Economics of rejuvenation of old pomegranate orchard have been clearly worked out in 3-4 consecutive years with the set particulars of operations like sale of fruit before rejuvenation, cost of hard core pruning, sale of woods, cost of intercropping, profits from intercrops, sale of fruits from rejuvenated trees and finally gross income have been found very encouraging with high benefit cost ratio with other added advantages. Economics of rejuvenation (tentative) of old pomegranate orchard is given in table 13.1.

Table 13.1: Economics of Rejuvenation of Old Pomegranate Orchard (One Hectare)

S. No.	Items/particulars of operations	Years			
		Year-I	Year-II	Year-III	Year-IV
1.	Sale of fruits before rejuvenation	25000.00	-	-	-
2.	Cost of Hardcore pruning	(-)10,000.00	-	-	-
3.	Cost of orchard floor cleaning	(-)04,000.00			
4.	Cost of inputs and application charges	(-)10,000.00			
5.	Sale of woods	10,000.00	-	-	-
6.	Cost of Inter-cropping	(-)25,000.00	(-)30,000.00	(-)20,000.00	(-)20,000.00
7.	Profit from Inter-crops	50,000.00	60,000.00	35,000.00	35,000.00
8.	Sale of fruits from rejuvenated trees	-		80,000.00	1,60,000.00
	GrossIncome	36,000.00	30,000.00	95,000.00	1,75,000.00

Impact

Hence, with the scientific skill and approach, complete transformation can be brought about by changing the exhausted phase of the old trees to behave and bear like young commercial trees with sustained performance. This transformation gave great impetus and boost to the enhanced quality fruit production having high economic returns. This wonder became possible only with the adoption of scientific rejuvenation technology. It is indeed an example, when the utmost transformation tried through rejuvenation pruning of old senile, unproductive orchard, without destroying the actual identity and fortunately, only in time gap of two to three years it started behaving like young commercial orchard with sustained economic returns. As a result, the increasing trend of production and productivity of quality yield in coming years as commercial bearing orchards, the fruit growers having old plantations giving uneconomical yield convincingly started adopting the technology developed by various Research Institutes/State Agricultural Universities. The success of rejuvenation technology in transforming the old orchards into young commercial bearing orchards have become the most important consideration in National Policies of Government Horticulture Developmental Plan.

Conclusion

Rejuvenation technology in fruit trees is visionary approach to save the old senile declining trees for future and sustainable quality fruit production. This technology is unique integration of invasive exhaustive operations and skillful management for bringing real desired transformation. The real and unique transformation is for potential and profitable exploitation. This technology adoption for old senile fruit trees is an ultimate option not only to save the declining plants/trees but also the effort to transform it from unproductive to productive ones by restoration of youthful vigour. The technology is strategic and innovative approach advocates the phase change through scientific skillful operation. The rejuvenation of old senile fruit trees, particularly pomegranate presents (a variety) many cross cutting issues and practical solution on the theme of restoration of youthful vigour, enhancement of quantum and quality fruit production, period of sustainability, economic feasibility. All these require further sound experimentation, implementation and success stories. Until now all but little existed in the way of survivability of declining.

14

Water Management in Pomegranate (*Punica granatum* L.)

S. D. Gorantiwar[1] and D.T. Meshram[2]

[1]Department of Irrigation and Drainage Engineering,
MPKV, Rahuri -413 722, Maharashtra
[2]ICAR-National Research Centre on Pomegranate, Solapur-413 255, Maharashtra
Email : sdgorantiwar@gmail.com

Preamble

There is a need to apply water according to water requirement of the crops. In Maharashtra pomegranate is predominately grown in the districts of Solapur, Ahmednagar, Pune, Nasik, Sangli, Satara and Osmanabad. Water is scarce commodity in these districts and hence need appropriate water management for pomegranate. The water requirement of pomegranate crop depends on age, season, location and management strategies.

The Government of Maharashtra encouraged the use of drip irrigation since 1986 by way of subsidy and in 1990 Government of India, with State Government, provided assistance to farmers for adopting the drip irrigation method to horticultural crops that has transformed the efficiency in water use effectively and as a result of which the pomegranate is being irrigated by drip irrigation method in the state of Maharashtra. On one side, on realizing the importance of drip irrigation method in the water scarcity situation, Government is encouraging its use and farmers are also adopting this method for pomegranate in a greater way; however on the other hand there is no specific information on the water requirement of pomegranate (WR) by drip irrigation.

This leads to inappropriate application of water to pomegranate, resulting in under or over irrigation and finally either in decreased productivity and/or increased wastage of scarce water resources. Therefore it is necessary to estimate the water requirement of pomegranate.

Computation of water requirement needs the measurement of evapotranspiration of the pomegranate (ETc). The ETc can be estimated by multiplying ETr with crop coefficient (Kc). Hence accurate estimation of ETr and Kc are of paramount importance for proper irrigation scheduling.

Crop Coefficient (KC) Values

It is important to know the values of Kc and ETr to estimate ETc and eventually water requirement of pomegranate (WR). Kc values of most of the crops are not available locally and hence in irrigation water management of these crops, the Kc values reported by FAO are often used. So is the case with pomegranate too. Hence the Kc values were locally determined. These are presented in tables 14.1 to 14.3 for pomegranate for different *Bahar*.

Table 14.1: Weekly Shaded Area (SA) and Crop Coefficient (Kc) Values of 1st to 5th year Pomegranate Trees for Mirg Bahar.

M W	Age of pomegranate tree (year)									
	1		2		3		4		5	
	SA	KC	SA	KC	SA	KC	SA	KC	SA	KC
31	0.06	0.16	0.10	0.22	0.03	0.13	0.04	0.14	0.05	0.15
32	0.06	0.16	0.11	0.24	0.06	0.16	0.09	0.20	0.10	0.22
33	0.06	0.17	0.13	0.25	0.09	0.21	0.13	0.26	0.15	0.29
34	0.06	0.17	0.13	0.27	0.14	0.26	0.17	0.32	0.21	0.37
35	0.07	0.17	0.15	0.29	0.18	0.33	0.21	0.38	0.26	0.44
36	0.07	0.18	0.16	0.31	0.23	0.40	0.27	0.46	0.30	0.50
37	0.07	0.18	0.17	0.32	0.27	0.46	0.32	0.52	0.36	0.59
38	0.08	0.19	0.19	0.34	0.32	0.53	0.39	0.63	0.42	0.67
39	0.08	0.19	0.20	0.36	0.40	0.64	0.45	0.71	0.47	0.74
40	0.08	0.19	0.21	0.38	0.45	0.71	0.50	0.78	0.53	0.82
41	0.08	0.2	0.23	0.40	0.51	0.79	0.56	0.87	0.57	0.88
42	0.09	0.2	0.24	0.41	0.56	0.86	0.62	0.95	0.62	0.95
43	0.09	0.2	0.26	0.45	0.61	0.93	0.67	1.02	0.67	1.02

Contd...

M W	Age of pomegranate tree (year)									
	1		2		3		4		5	
	SA	KC	SA	KC	SA	KC	SA	KC	SA	KC
44	0.09	0.21	0.28	0.47	0.66	1.01	0.73	1.10	0.73	1.11
45	0.09	0.21	0.30	0.49	0.71	1.07	0.76	1.15	0.78	1.17
46	0.09	0.21	0.30	0.50	0.71	1.07	0.76	1.14	0.78	1.17
47	0.10	0.21	0.30	0.50	0.71	1.07	0.76	1.15	0.78	1.18
48	0.10	0.22	0.30	0.50	0.71	1.07	0.77	1.15	0.78	1.18
49	0.10	0.22	0.30	0.51	0.71	1.07	0.77	1.16	0.78	1.18
50	0.10	0.22	0.31	0.51	0.71	1.08	0.77	1.16	0.79	1.18
51	0.10	0.22	0.31	0.51	0.71	1.08	0.77	1.17	0.79	1.18
52	0.10	0.22	0.31	0.51	0.72	1.08	0.78	1.17	0.79	1.18
1	0.11	0.23	0.28	0.47	0.66	1.00	0.74	1.12	0.76	1.14
2	0.11	0.23	0.27	0.45	0.63	0.96	0.72	1.09	0.74	1.11
3	0.11	0.23	0.25	0.44	0.60	0.92	0.69	1.05	0.72	1.08
4	0.11	0.24	0.24	0.42	0.59	0.90	0.67	1.02	0.70	1.06
5	0.12	0.24	0.23	0.40	0.57	0.88	0.66	1.01	0.69	1.05
6	0.12	0.25	0.21	0.38	0.56	0.86	0.63	0.97	0.68	1.03
7	0.12	0.25	0.20	0.36	0.54	0.86	0.61	0.93	0.67	1.01
8	0.12	0.25	0.19	0.35	0.52	0.83	0.60	0.91	0.65	0.99
9	0.13	0.26	0.18	0.33	0.50	0.80	0.57	0.88	0.64	0.98
10	0.13	0.26	0.17	0.31	0.49	0.78	0.56	0.86	0.63	0.96
11	0.13	0.27	0.15	0.29	0.47	0.76	0.54	0.83	0.60	0.92
12	0.11	0.23	0.14	0.28	0.45	0.74	0.46	0.73	0.53	0.82
13	0.11	0.23	0.14	0.28	0.40	0.64	0.47	0.73	0.53	0.83
14	0.11	0.23	0.14	0.28	0.41	0.65	0.47	0.74	0.53	0.83
15	0.11	0.23	0.15	0.28	0.41	0.65	0.47	0.74	0.54	0.83
16	0.11	0.23	0.15	0.29	0.41	0.65	0.48	0.75	0.54	0.83
17	0.11	0.23	0.15	0.29	0.41	0.66	0.48	0.75	0.54	0.83
18	0.11	0.24	0.15	0.30	0.42	0.66	0.48	0.75	0.54	0.84

Contd...

M W	Age of pomegranate tree (year)									
	1		2		3		4		5	
	SA	KC	SA	KC	SA	KC	SA	KC	SA	KC
19	0.11	0.24	0.16	0.30	0.42	0.66	0.48	0.75	0.56	0.86
20	0.12	0.24	0.16	0.30	0.42	0.67	0.49	0.76	0.56	0.86
21	0.12	0.24	0.16	0.31	0.43	0.67	0.49	0.76	0.56	0.87
22	0.12	0.25	0.16	0.31	0.43	0.68	0.49	0.77	0.57	0.87
23	0.12	0.25	0.17	0.32	0.43	0.68	0.49	0.77	0.57	0.88
24	0.12	0.25	0.17	0.32	0.44	0.69	0.50	0.78	0.57	0.88
25	0.12	0.25	0.17	0.32	0.44	0.70	0.50	0.78	0.58	0.89
26	0.13	0.25	0.18	0.33	0.45	0.70	0.50	0.78	0.58	0.89
27	0.13	0.26	0.18	0.33	0.45	0.71	0.51	0.79	0.59	0.90
28	0.13	0.26	0.18	0.34	0.45	0.71	0.51	0.79	0.59	0.91
29	0.13	0.27	0.19	0.34	0.46	0.72	0.51	0.79	0.60	0.91
30	0.14	0.27	0.19	0.35	0.46	0.72	0.51	0.80	0.60	0.92

Table 14.2: Weekly Shaded Area (SA) and Crop Coefficient (Kc) Values of 1st to 5th Year Pomegranate Trees for *Ambhe Bahar*.

M W	Age of pomegranate tree (year)									
	1		2		3		4		5	
	SA	KC	SA	KC	SA	KC	SA	KC	SA	KC
1	0.06	0.16	0.10	0.22	0.03	0.13	0.04	0.14	0.05	0.15
2	0.06	0.16	0.11	0.24	0.06	0.16	0.09	0.20	0.10	0.22
3	0.06	0.17	0.13	0.25	0.09	0.21	0.13	0.26	0.15	0.29
4	0.06	0.17	0.13	0.27	0.14	0.26	0.17	0.32	0.21	0.37
5	0.07	0.17	0.15	0.29	0.18	0.33	0.21	0.38	0.26	0.44
6	0.07	0.18	0.16	0.31	0.23	0.40	0.27	0.46	0.30	0.50
7	0.07	0.18	0.17	0.32	0.27	0.46	0.32	0.52	0.36	0.59
8	0.08	0.19	0.19	0.34	0.32	0.53	0.39	0.63	0.42	0.67
9	0.08	0.19	0.20	0.36	0.40	0.64	0.45	0.71	0.47	0.74
10	0.08	0.19	0.21	0.38	0.45	0.71	0.50	0.78	0.53	0.82
11	0.08	0.2	0.23	0.40	0.51	0.79	0.56	0.87	0.57	0.88
12	0.09	0.2	0.24	0.41	0.56	0.86	0.62	0.95	0.62	0.95

Contd...

M W	Age of pomegranate tree (year)									
	1		2		3		4		5	
	SA	KC	SA	KC	SA	KC	SA	KC	SA	KC
13	0.09	0.2	0.26	0.45	0.61	0.93	0.67	1.02	0.67	1.02
14	0.09	0.21	0.28	0.47	0.66	1.01	0.73	1.10	0.73	1.11
15	0.09	0.21	0.30	0.49	0.71	1.07	0.76	1.15	0.78	1.17
16	0.09	0.21	0.30	0.50	0.71	1.07	0.76	1.14	0.78	1.17
17	0.10	0.21	0.30	0.50	0.71	1.07	0.76	1.15	0.78	1.18
18	0.10	0.22	0.30	0.50	0.71	1.07	0.77	1.15	0.78	1.18
19	0.10	0.22	0.30	0.51	0.71	1.07	0.77	1.16	0.78	1.18
20	0.10	0.22	0.31	0.51	0.71	1.08	0.77	1.16	0.79	1.18
21	0.10	0.22	0.31	0.51	0.71	1.08	0.77	1.17	0.79	1.18
22	0.10	0.22	0.31	0.51	0.72	1.08	0.78	1.17	0.79	1.18
23	0.11	0.23	0.28	0.47	0.66	1.00	0.74	1.12	0.76	1.14
24	0.11	0.23	0.27	0.45	0.63	0.96	0.72	1.09	0.74	1.11
25	0.11	0.23	0.25	0.44	0.60	0.92	0.69	1.05	0.72	1.08
26	0.11	0.24	0.24	0.42	0.59	0.90	0.67	1.02	0.70	1.06
27	0.12	0.24	0.23	0.40	0.57	0.88	0.66	1.01	0.69	1.05
28	0.12	0.25	0.21	0.38	0.56	0.86	0.63	0.97	0.68	1.03
29	0.12	0.25	0.20	0.36	0.54	0.86	0.61	0.93	0.67	1.01
30	0.12	0.25	0.19	0.35	0.52	0.83	0.60	0.91	0.65	0.99
31	0.13	0.26	0.18	0.33	0.50	0.80	0.57	0.88	0.64	0.98
32	0.13	0.26	0.17	0.31	0.49	0.78	0.56	0.86	0.63	0.96
33	0.13	0.27	0.15	0.29	0.47	0.76	0.54	0.83	0.60	0.92
34	0.11	0.23	0.14	0.28	0.45	0.74	0.46	0.73	0.53	0.82
35	0.11	0.23	0.14	0.28	0.40	0.64	0.47	0.73	0.53	0.83
36	0.11	0.23	0.14	0.28	0.41	0.65	0.47	0.74	0.53	0.83
37	0.11	0.23	0.15	0.28	0.41	0.65	0.47	0.74	0.54	0.83
38	0.11	0.23	0.15	0.29	0.41	0.65	0.48	0.75	0.54	0.83
39	0.11	0.23	0.15	0.29	0.41	0.66	0.48	0.75	0.54	0.83
40	0.11	0.24	0.15	0.30	0.42	0.66	0.48	0.75	0.54	0.84
41	0.11	0.24	0.16	0.30	0.42	0.66	0.48	0.75	0.56	0.86
42	0.12	0.24	0.16	0.30	0.42	0.67	0.49	0.76	0.56	0.86
43	0.12	0.24	0.16	0.31	0.43	0.67	0.49	0.76	0.56	0.87
44	0.12	0.25	0.16	0.31	0.43	0.68	0.49	0.77	0.57	0.87
45	0.12	0.25	0.17	0.32	0.43	0.68	0.49	0.77	0.57	0.88

Contd...

M W	Age of pomegranate tree (year)									
	1		2		3		4		5	
	SA	KC	SA	KC	SA	KC	SA	KC	SA	KC
46	0.12	0.25	0.17	0.32	0.44	0.69	0.50	0.78	0.57	0.88
47	0.12	0.25	0.17	0.32	0.44	0.70	0.50	0.78	0.58	0.89
48	0.13	0.25	0.18	0.33	0.45	0.70	0.50	0.78	0.58	0.89
49	0.13	0.26	0.18	0.33	0.45	0.71	0.51	0.79	0.59	0.90
50	0.13	0.26	0.18	0.34	0.45	0.71	0.51	0.79	0.59	0.91
51	0.13	0.27	0.19	0.34	0.46	0.72	0.51	0.79	0.60	0.91
52	0.14	0.27	0.19	0.35	0.46	0.72	0.51	0.80	0.60	0.92

Table 14.3: Weekly Shaded Area (SA) and Crop Coefficient (Kc) Values of 1st to 5th Year Pomegranate Trees for *Hasta Bahar*.

M W	Age of pomegranate tree (year)									
	1		2		3		4		5	
	SA	KC	SA	KC	SA	KC	SA	KC	SA	KC
36	0.06	0.16	0.10	0.22	0.03	0.13	0.04	0.14	0.05	0.15
37	0.06	0.16	0.11	0.24	0.06	0.16	0.09	0.20	0.10	0.22
38	0.06	0.17	0.13	0.25	0.09	0.21	0.13	0.26	0.15	0.29
39	0.06	0.17	0.13	0.27	0.14	0.26	0.17	0.32	0.21	0.37
40	0.07	0.17	0.15	0.29	0.18	0.33	0.21	0.38	0.26	0.44
41	0.07	0.18	0.16	0.31	0.23	0.40	0.27	0.46	0.30	0.50
42	0.07	0.18	0.17	0.32	0.27	0.46	0.32	0.52	0.36	0.59
43	0.08	0.19	0.19	0.34	0.32	0.53	0.39	0.63	0.42	0.67
44	0.08	0.19	0.20	0.36	0.40	0.64	0.45	0.71	0.47	0.74
45	0.08	0.19	0.21	0.38	0.45	0.71	0.50	0.78	0.53	0.82
46	0.08	0.2	0.23	0.40	0.51	0.79	0.56	0.87	0.57	0.88
47	0.09	0.2	0.24	0.41	0.56	0.86	0.62	0.95	0.62	0.95
48	0.09	0.2	0.26	0.45	0.61	0.93	0.67	1.02	0.67	1.02
49	0.09	0.21	0.28	0.47	0.66	1.01	0.73	1.10	0.73	1.11
50	0.09	0.21	0.30	0.49	0.71	1.07	0.76	1.15	0.78	1.17
51	0.09	0.21	0.30	0.50	0.71	1.07	0.76	1.14	0.78	1.17
52	0.10	0.21	0.30	0.50	0.71	1.07	0.76	1.15	0.78	1.18
1	0.10	0.22	0.30	0.50	0.71	1.07	0.77	1.15	0.78	1.18
2	0.10	0.22	0.30	0.51	0.71	1.07	0.77	1.16	0.78	1.18
3	0.10	0.22	0.31	0.51	0.71	1.08	0.77	1.16	0.79	1.18

Contd...

M W	Age of pomegranate tree (year)									
	1		2		3		4		5	
	SA	KC	SA	KC	SA	KC	SA	KC	SA	KC
4	0.10	0.22	0.31	0.51	0.71	1.08	0.77	1.17	0.79	1.18
5	0.10	0.22	0.31	0.51	0.72	1.08	0.78	1.17	0.79	1.18
6	0.11	0.23	0.28	0.47	0.66	1.00	0.74	1.12	0.76	1.14
7	0.11	0.23	0.27	0.45	0.63	0.96	0.72	1.09	0.74	1.11
8	0.11	0.23	0.25	0.44	0.60	0.92	0.69	1.05	0.72	1.08
9	0.11	0.24	0.24	0.42	0.59	0.90	0.67	1.02	0.70	1.06
10	0.12	0.24	0.23	0.40	0.57	0.88	0.66	1.01	0.69	1.05
11	0.12	0.25	0.21	0.38	0.56	0.86	0.63	0.97	0.68	1.03
12	0.12	0.25	0.20	0.36	0.54	0.86	0.61	0.93	0.67	1.01
13	0.12	0.25	0.19	0.35	0.52	0.83	0.60	0.91	0.65	0.99
14	0.13	0.26	0.18	0.33	0.50	0.80	0.57	0.88	0.64	0.98
15	0.13	0.26	0.17	0.31	0.49	0.78	0.56	0.86	0.63	0.96
16	0.13	0.27	0.15	0.29	0.47	0.76	0.54	0.83	0.60	0.92
17	0.11	0.23	0.14	0.28	0.45	0.74	0.46	0.73	0.53	0.82
18	0.11	0.23	0.14	0.28	0.40	0.64	0.47	0.73	0.53	0.83
19	0.11	0.23	0.14	0.28	0.41	0.65	0.47	0.74	0.53	0.83
20	0.11	0.23	0.15	0.28	0.41	0.65	0.47	0.74	0.54	0.83
21	0.11	0.23	0.15	0.29	0.41	0.65	0.48	0.75	0.54	0.83
22	0.11	0.23	0.15	0.29	0.41	0.66	0.48	0.75	0.54	0.83
23	0.11	0.24	0.15	0.30	0.42	0.66	0.48	0.75	0.54	0.84
24	0.11	0.24	0.16	0.30	0.42	0.66	0.48	0.75	0.56	0.86
25	0.12	0.24	0.16	0.30	0.42	0.67	0.49	0.76	0.56	0.86
26	0.12	0.24	0.16	0.31	0.43	0.67	0.49	0.76	0.56	0.87
27	0.12	0.25	0.16	0.31	0.43	0.68	0.49	0.77	0.57	0.87
28	0.12	0.25	0.17	0.32	0.43	0.68	0.49	0.77	0.57	0.88
29	0.12	0.25	0.17	0.32	0.44	0.69	0.50	0.78	0.57	0.88
30	0.12	0.25	0.17	0.32	0.44	0.70	0.50	0.78	0.58	0.89
31	0.13	0.25	0.18	0.33	0.45	0.70	0.50	0.78	0.58	0.89
32	0.13	0.26	0.18	0.33	0.45	0.71	0.51	0.79	0.59	0.90
33	0.13	0.26	0.18	0.34	0.45	0.71	0.51	0.79	0.59	0.91
34	0.13	0.27	0.19	0.34	0.46	0.72	0.51	0.79	0.60	0.91
35	0.14	0.27	0.19	0.35	0.46	0.72	0.51	0.80	0.60	0.92

Estimation of ETr

Doorenbos and Pruitt (1977) have defined the term reference crop evapotranspiration (ET_r) to avoid ambiguity involved in the interpretation of evapotranspiration as evapotranspiration of well watered, actively growing green grass which is clipped to uniform height 8-15 cm completely shading the soil, not short of water and covering an extensive area. Values measured or calculated at different locations or in different seasons are therefore comparable as they refer to the evapotranspiration from same reference surface.

There are many methods reported in the literature to estimate reference crop evapotranspiration. The important methods include: Penman, Modified Penman, Penman Monteith, Hargreaves-Samani, FAO Pan Evaporation, Blanney-Criddle, FAO Radiation, Jensen-Haise, Priestly-Taylor, Thronthwaite, Christiansen. The excellent reviews of these methods have been provided by Michael (2008) and Doorenbos and Pruitt (1977). Many researchers have compared different evapotranspiration methods. Patil and Gorantiwar (2009) have provided the detail review on the comparison of different methods. The Penman-Monteith method was recommended by FAO as the most appropriate method of determining reference crop evapotranspiration.

It is proposed to use the Penman-Monteith method (equation 1) for the estimation of ETr as this method is widely used, recommended by FAO; and many researchers found this method close to the actual measurements of ETr compared to other methods (Patil and Gorantiwar, 2009).

$$ET_r = \frac{0.408\Delta\, R_n - G + \gamma\, \dfrac{900}{T+273}\, u_{2\,e_s - e_a}}{\Delta + \gamma\, 1 + 0.34u_2}$$

where,

ET_r = reference evapotranspiration (mm/day)
G = soil heat flux density (MJ/m²/day)
R_n = net radiation (MJ/m²/day)
T = mean daily air temperature (⁰C)
γ = psychometric constant (kPa/⁰C)
Δ = slope of saturation vapour pressure function (kPa/⁰C)
e_s = saturation vapour pressure at air temperature T (kPa)
e_a = actual vapour pressure at dew point temperature (kPa)
u_2 = average daily wind speed at 2 m height (m/sec)

The details of the computation of different parameters of the equation (1) can be found in Smith *et al* (1991). As it was huge task to estimate the daily ETr by equation (1), the user friendly computer software *"Phule Jal"* was prepared by Mahatma Phule Krishi Vidyapeeth, Rahuri to estimate reference crop evapotranspiration by using Penman Monteith method.

This method needs the daily values of meteorological parameters viz. maximum temperature, minimum temperature, maximum relative humidity, minimum relative humidity, wind speed and sunshine hours. The daily records of these parameters were obtained from the Indian Meteorological Department, Pune and Dryland Agricultural Research Center, Solapur and the monthly averages of these parameters are presented in table 14.4.

Table 14.4: Average Weekly ETr (mm) Values for Major Pomegranate Growing Districts of Maharashtra

MW	Station							MW	Station						
	Sol	Ahm	Pun	Nas	Sat	San	Osm		Sol	Ahm	Pun	Nas	Sat	San	Osm
1	25.2	21.1	18.1	20.4	27.4	19.0	23.2	27	34.2	33.0	24.6	26.1	34.8	25.7	36.0
2	25.4	21.9	19.1	21.4	27.7	20.8	23.9	28	31.9	31.3	24.6	24.8	34.9	27.1	33.7
3	26.6	23.3	20.5	23.0	28.9	21.7	25.6	29	30.3	30.5	22.2	23.0	30.8	23.8	33.0
4	28.5	24.7	21.7	24.1	32.0	22.8	27.0	30	29.3	29.4	21.9	24.1	29.0	24.4	30.6
5	30.1	26.1	22.3	26.7	32.1	22.9	28.4	31	30.0	29.3	21.7	23.3	29.9	25.5	29.4
6	31.8	27.9	24.7	28.8	33.6	25.4	30.0	32	28.7	28.5	19.8	21.4	28.3	22.0	29.5
7	33.0	29.6	27.1	31.0	35.5	27.6	32.1	33	29.1	29.7	21.4	22.5	29.3	22.1	31.3
8	36.1	31.4	28.5	31.5	40.4	28.4	34.1	34	29.2	29.8	21.0	21.1	33.2	20.2	32.5
9	38.8	34.9	29.8	34.4	40.2	31.0	37.6	35	31.0	28.5	21.7	22.7	34.7	22.0	31.1
10	39.3	35.2	31.7	37.1	41.9	32.0	38.2	36	30.3	30.2	24.3	24.5	31.3	26.2	32.6
11	40.5	36.8	34.0	39.4	40.8	35.4	39.5	37	30.5	29.3	24.2	24.3	30.2	26.4	30.9
12	42.7	39.3	36.5	43.6	43.9	37.1	42.8	38	29.3	28.6	24.3	25.6	28.9	26.0	29.7
13	44.6	41.4	38.0	44.1	46.5	37.8	45.1	39	28.2	27.2	23.0	25.1	31.1	23.8	28.6
14	45.3	42.2	40.0	46.3	48.1	40.1	45.9	40	28.2	27.8	22.6	25.8	28.3	22.3	29.0
15	47.1	44.5	41.1	48.3	50.1	41.9	47.8	41	28.2	28.7	22.7	25.3	31.7	24.0	31.6
16	50.2	47.8	42.0	50.0	52.2	42.4	52.1	42	29.0	28.0	23.2	26.4	31.1	23.4	30.3
17	51.0	50.0	44.5	52.8	51.3	46.2	50.7	43	29.2	27.0	23.2	26.3	32.8	24.1	28.7
18	52.8	52.9	46.0	52.6	53.1	48.6	54.3	44	28.1	25.9	22.4	25.1	29.2	24.3	27.8
19	54.4	50.8	46.2	51.1	54.9	44.4	54.3	45	28.1	24.7	21.8	24.0	29.8	23.1	26.6
20	56.0	53.6	46.2	52.0	56.3	44.6	58.3	46	27.5	23.6	20.5	23.6	32.6	21.0	25.3
21	53.7	52.9	44.4	49.0	54.7	43.0	56.6	47	25.9	22.4	19.7	22.2	27.5	21.6	24.1
22	51.5	49.8	40.3	45.9	48.7	42.7	53.1	48	25.7	22.0	18.8	21.3	28.8	20.7	23.4
23	45.1	44.2	38.6	43.1	45.3	38.8	47.1	49	25.0	21.5	18.6	20.4	28.4	21.0	22.9
24	38.9	37.6	29.4	34.7	39.3	29.7	39.7	50	24.4	21.0	18.5	20.4	25.8	18.4	22.9
25	37.5	36.1	27.7	30.9	39.5	32.2	38.8	51	24.8	20.6	18.8	20.9	26.9	19.9	22.4
26	34.1	34.8	24.1	27.9	28.1	25.0	37.3	52	27.5	23.3	20.8	22.9	28.0	22.2	25.2

| | Average Annual ET r(mm) | 1803 | 1692 | 1428 | 1602 | 1879 | 1482 | 1812 |

(Note: MW-Standard Meteorology Week, Sol –Solapur, Ahm- Ahmednagar, Pun – Pune Nas-Nasik, Sat –Satara, San-Sangli and Osm – Osmanabad)

Estimation of ETc

The weekly values of ETr and Kc were used to obtain weekly values of ETc by equation (2) for *Ambhe, Mirg* and *Hasta bahars* seasons for all the years and stations.

$$ET_c = ET_r * k_c \qquad\qquad (2)$$

where

ET_c = pomegranate evapotranspiration (mm/day)

ET_r = reference crop evapotranspiration. (mm/day)

K_c = crop coefficient of pomegranate.

Water Requirement

The water requirement by the surface irrigation methods is equal to the crop evapotranspiration estimated by the equation (2). However water requirement by the drip irrigation method is less than the water requirement of the surface irrigation methods as in drip irrigation method unlike in surface irrigation method, it is possible to apply water to the effective root zone only. Hence water requirement in case of drip irrigation method is estimated by the equation (3).

$$WR = ETc * Fa \qquad\qquad (3)$$

where

WR = water requirement (mm/day)

Fa = Area factor (fraction)

Area Factor

Area factor is the proportion of the effective root zone with respect to the total area. The area factor hence varies with the crop growth period and the age of the crop. In general it has been reported in the literature that for most of the deciduous crop, the effective root zone area below the soil surface is the area occupied by the canopy above the soil. The canopy area is the shaded area at solar noon hour. The weekly values of shaded area for pomegranate are presented in Tables 1 to 3.

Water to be Applied

The farmers need the information on water to be applied to each pomegranate tree. Water to be applied was estimated on weekly basis for the pomegranate trees up to the age of 5 by using the equation (4).

$$WA = WR * A/eff \qquad\qquad (4)$$

where

WA = water to be applied to each tree (liter/day)
A = area occupied by each tree (m²)
eff = efficiency of the drip irrigation system (fraction)

The water to be applied to the pomegranate plantation irrigated by surface irrigation methods can be calculated by using the ETr values (Table 14.4) and Kc values (Tables 14.1 to 14.3), if the efficiency of the surface irrigation method is known (equation 2). The water to be applied to the pomegranate plantation irrigated by drip irrigation method can be calculated by using the ETr values (Table 4) and Kc and SA values (Tables 14.1 to 14.3), if the efficiency of the drip irrigation method and the area covered by the pomegranate tree are known.

Usually the pomegranate is spaced at 4.5 x 3 m and the drip irrigation systems are designed for 90% efficiency. Hence the values of water to be applied to pomegranate of different seasons for different stations can be estimated for the tree spacing of 4.5 x 3.0 m (or other spacing) and drip irrigation efficiency of 90 % (or other efficiency) (equation 4) by using the information on crop coefficient (Tables 14.1 to 14.3) and reference crop evapotranspiration values (Table 14.4).

15

Present Status, Scope and Strategies of Pomegranate Cultivation under Agro-climatic Conditions of Chhattisgarh

S.S. Shaw, Vijay Kumar and R.O. Das

Indira Gandhi Agricultural University, Raipur-492 012, Chhattisgarh

Introduction

Pomegranate belongs to genera *Punica* and family Lytheraceae (Joshi, 1956) and is an important semi-arid fruit, cultivated commercially in over 1.93 lakhs hectares area in western India. It is a rich source of carbohydrate (14.5%), protein (1.6%), calcium (10 mg/100g), phosphorus (70mg/ 100g), iron (0.3 mg/100g) and vitamin C (65mg/100g). Pomegranate is a rich source of nutrients packed with medicinal properties and it is necessary to enhance its consumption to overcome malnutrition among rural and tribal people. Therefore, it is important to determine its potential for its commercial exploitation and expansion in Chhattisgarh state. In India, it is considered as a crop of the arid and semi-arid regions because it withstands different soil and climatic stresses (Kaulgud, 2001). Under temperate climate, pomegranate behaves as deciduous but in sub-tropical and tropical climate it is an evergreen or partially deciduous. It is emerging as one of the important fruit crops owing to hardiness and ability to withstand adverse soil and climatic conditions. The main pomegranate producing areas are distributed between 300 m to 950 m above mean sea level in hot arid and semi-arid regions having tropical and subtropical climate. In the present paper, information pertaining to Chhattisgarh and its climatic and soil conditions and future scope and potential of pomegranate has been envisaged.

There is an urgent need for crop diversification in Chhatisgarh state for sustainable agriculture since the production of food grains is not much profitable for small and marginal farmers. The average productivity of some important crops in the three agro-climatic zones varies from 1.0 to 1.1 t/ha for rain fed rice, 1.6 to 1.8 t/ha for irrigated rice,1.4 to 1.5 t/ha for wheat,1.4 to 1.5 t/ha for maize, 5.6 to 6.3 for rape-mustard and 0.5 to 0.6 t/ha for gram. It can clearly be seen that the productivity of not only rice but also of other crops is low, hence the farmers are unable to obtain economic benefits from agriculture and it has remained as subsistence agriculture till now. Moreover, frequent drought and erratic rainfall affect the crop yield adversely. Introduction of rain fed horticulture in the drought prone upland areas can help the farmers in getting the income for their livelihoood. Owing to diversities existing in soil, climate and edaphic factors, there is scope of cultivation of all kinds of fruits and vegetables of tropical, subtropical and temperate climate in the state.Most of the fruit crops can successfully be cultivated as the dry land fruit crops and among these pomegranate needs special mention on account of its nutritional and medicinal importance and wider adaptability to climatic conditions. Inter-cropping with low growing vegetables, pulses or green manure crops is beneficial in pomegranate orchards. In arid regions, inter-cropping is possible only during the rainy season, whereas winter vegetables are feasible in irrigated areas.

Present Status and Scope

India ranks first in the world with respect to pomegranate area (0.125 million ha) and production (1.14 million tones). Maharashtra contributes more than 75% of the total area alone followed by Karnataka and Andhra Pradesh (Ram Chandra *et al.*, 2010) where good quality fruits are produced due to dry and hot climate. It is cultivated to a large extent in the northern dry districts of Karnataka state. India contributes 40-45 % of world pomegranate production. With the increase in population, domestic demand for fruit has increased substantially. The fruit is fetching much foreign exchange for the country as sizeable quantity of fruits is being exported from these states. India exports around 32 thousand tonnes of fruits valued at Rs 4160 millions. In Chhattisgarh, mango, guava, papaya, bananas Jackfruit, custard apple are major fruit crops cultivated in an area of around 58,441 hectares (Anon, 2002) under different soil and agro-climatic conditions and currently there is negligible area(100 hectares) and production(40 tones) under pomegranate cultivation(personal communication, 2014).

Chhattisgarh state has about 6 lakh hectares wasteland and under dry land horticulture,there is good scope for cultivation of pomegranate for utilization of upland areas having poor soil conditions (Anon.,2002). Since some of the fruits like pomegranate, ber and aonla are particularly sensitive to humidity, rainfall and temperatures, hence these fruit crops are more suitable for cultivation under dry regions that receive less than 1100 mm of total rainfall because atmospheric humidity requirement of these fruit crops is as low as 50%. Pomegranate thrives well under hot, dry summer and cold winter provided irrigation facilities are available. The tree requires hot and dry climate during fruit development and ripening. It can tolerate frost to a considerable extent in dormant stage, but is injured at temperature below -11⁰C.

Atmospheric humidity influences at various stages of growth, flowering, fruit set, maturity and humidity also influence disease and pest occurrence in pomegranate. One of the major constraints of pomegranate cultivation in Chhattisgarh could probably be the sub- humid climate under which fruits are severely damaged by pomegranate butterfly and do not develop sweetness. Moreover, high humidity during flowering and fruiting may promote disease like powdery mildew. As per the data base prepared on the basis of GIS analysis (Anon., 2002), in Chhattisgarh a total area of 1 lakh hectares has been found suitable for cultivation of specific fruit crops including pomegranate, grape and ber which are considered to be adaptable to local climatic parameters and out of which most of the area lies in southern part of Rajnandgaon (39,678 hectares) and Durg (68,246) districts.

Soils

Pomegranate is adapted to a wide range of soils ranging from acidic sandy loam to alkaline calcareous soils and everything in between. In India, they have been known to grow in rocky gravel soils with organic substrate. The only soil that will not support pomegranates is heavy clay, because soils with excessive clay tend to present drainage problems. However, best results are obtained in deep heavy loam and well drained soils. It can be successfully cultivated in soils with a depth greater than 60 cm, slope 2 to 15 % and pH 6 to 8. Neutral to slightly acidic soil is best for pomegranate. They tolerate hot windy exposed conditions & slopes. It is sensitive to soil moisture fluctuations causing fruit cracking which is a serious problem of this crop. Pomegranate can withstand significant periods of drought, however fruiting can be diminished in drought conditions, so watering may be helpful for increasing its fruit yield.

The soils of Chhattisgarh vary considerably in the three agro-climatic zones with different nomenclature. The first two categories of the soils in the three agroclimatic zones are Lateritic (popularly known as Bhata) and Sandy loam (popularly known as Matasi) for Chhattisgarh Plains, coarse sandy (popularly known as Marhan) and sandy (popularly known as Tikra) for Bastar Plateau and Hilly soils and Tikra (popular name) for Northern hills which are very light type of soils with very low water retentive capacity. As a result water stress or drought conditions occur either during the crop growing season when there is a break of monsoon for more than 5-7 days or immediately after the withdrawal of monsoon. Therefore, there is an urgent need to diversify cropping pattern wherever conditions are favourable to grow horticultural crops and to earn higher net profit in unit area. Since, above mentioned categories of soils of Chhattisgarh zones are not so deep and pomegranate being shallow rooted can successfully be cultivated under such nature and kind of soil conditions. Pomegranates can also grow in relatively poor soil because the plants are shallow rooted & nutrients can be absorbed quickly. As with all fruit trees the addition of organic compost to the soil makes a big difference on improving the quantity & quality of fruit.

Geographical Features and Climate

Geographically, Chhattisgarh is divided into three distinct land areas *viz.*,Chhattisgarh Plains, Bastar Plateau and Northern Hill Zones. In the north of the state are the mighty Satpura Ranges, in the center the plains of River Mahanadi and its tributaries and in the south is the plateau of Bastar. The state receives annual rainfall ranging from less than 1200 mm to greater than 1600 mm in different areas. The border of Chhattisgarh is touched by the states Uttar Pradesh in the North, Bihar in the North East, Orissa in the East, Andhra Pradesh in the South and South East, Maharashtra inSouth West and Madhya Pradesh in the West. The general climate of Chhattisgarh state is dry sub-humid type where the annual potential evapo-transpiration is slightly higher than the annual rainfall. The average annual rainfall of the region is around 1400 season (June-October). The monsoon sets in around 10 June in the tip of the Bastar area and covers the entire area by 25th June. July and August are the wettest months. Rainfall in October month occurs due to cyclonic activity in the Bay of Bengal and October rainfall is most crucial for the productivity of rice in the state. Winter conditions set in from mid November when the average minimum temperature starts falling below 15°C. The northern districts especially Bilaspur division have more severe and longer winter period as compared to southern parts especially Bastar division. The atmospheric humidity is very high (>90%) during monsoon months and starts decreasing from October onwards and reaches as low as 15-20 percent during peak summer months. In view of prevailing climatic conditions, the area under pomegranate may be developed in Chhattisgarh around North western part of the state and western area adjoining the boundaries of M. P. and Maharastra states.

Strategies

1. Cultivars/varieties

Main focus should be on identification of varieties by using molecular characterization through DNA and isozyme markers. The relative effectiveness of different cultivars and training systems in different agro-climatic regions of Chhattisgarh need to be ascertained. The emphasis should be laid on development of varieties having high sugar (16%) and less acid (0.5%) and selection of varieties bearing high %age of fertile flowers. The development of varieties by hybridization having medium sized fruits, deep red to deep pink flesh, soft seeds and quite sweet juice is the need of hour. Breeding anardana type varieties with dark red and bold arils, high acidity, bigger fruit size and high yield needs special attention.

2. Crop Regulation

The pomegranate starts fruiting about 4 years after planting and continues for about 25 to 30 years. To regulate flowering, water is withheld for about two months in advance of the normal flowering season. A full grown pomegranate has tendency to bear flowers and fruits throughout the year. Crop regulation practices for taking *ambe bahar, mrig bahar and hasta bahar* need to be standardized with reference to time of inducing rest period to plant and yield and quality attributes of fruits under Chhattisgarh Agro-climatic conditions. To maintain productivity of the plants, generally one *bahar* fruiting is regulated, which depends upon market factors and availability of water. The feasibility of taking *bahar* crops should be explored depending upon the water availability for critical stage of plant growth under a particular *bahar* treatment. Studies have showed that *Ambe bahar* may be taken in the areas where enough water is available for better returns. Processing facilities also need to be created in the state for value addition.

Depending on patterns of precipitation, flowering can be induced during June-July (*mrig bahar*), September-October (*hasta bahar*) and January-February (*ambe bahar*). In areas having assured rainfall where precipitation is normally received in June and continues upto September, flowering in June is advantageous. Areas having assured irrigation potential during April- May, flowering during January can be taken and where monsoon starts early and withdraws by September induction of flowering in October is possible. Considering comparable yields, prices and irrigation needs it is recommended that October cropping could be substituted for January flowering and this needs further investigation with reference to Chhatisgarh region. Scope of *mrigbahar* may be explored in the Chhattisgarh region where there is scanty availability of irrigation water. Therefore, efforts should be directed to avail of rainy flowering season(*Mrig- bahar*) so that fruiting period coincides with the time of maximum water availability in the soil and the crop is taken without irrigation. Observations should be made on the bacterial blight incidence under three bahar treatments.

3. High Density Planting

Several studies have indicated success of high density planting of pomegranate in different growing regions. High density planting with less spacing gives 2-2.5 times more yield than that obtained when the normal planting distance of 5 X 5 m. is adopted closer spacing increases disease and pest incidence. Evaluation, conservation and cataloging of both exotic and indigenous pomegranate cultivars with especially compact, dwarf and thorn less types is needed to be done for Chhattisgarh agro-climatic conditions so that they can be brought under high density program in this state. There is need for evaluation and development of cultivation of tolerant varieties and insect –pest and disease scenario should be studied in relation to high density planting.

4. Agri-horticulture System

Since the pomegranate is unable to tolerate water stagnation and to avoid water stagnation problem during monsoon, raised and sunken bed technique has been found quite suitable for agro forestry practices in highly alkaline soil. *Punica granatum* was successfully grown on raised beds to avoid water stagnation and on sunken beds were constructed for the purpose of rice/wheat rotation in highly alkaline soils in the studies at CSSRI, Karnal (Dagar *et al.*, 2001). Similar kind of raised and sunken bed technologies may be standardized for soil and agro-climatic conditions of Chhattisgarh state. Looking to this the upland or midland areas of Chhattisgarh may be brought under pomegranate expansion. Pomegranate may be planted on the bunds of rice, wheat, vegetable or cereal crops in low lands. The studies may be undertaken with regards to interaction of field crops and pomegranate in relation to micro climate, growth and yields of component crops and physiology of bund planted pomegranate trees under the influence of crop management practices particularly with reference to irrigation to field crops. Under this system the pruning practices be standardized for fruit yield and quality.

5. Biotic and Abiotic Stress

In India, bacterial blight, wilt, fruit borer, thrips, sun scald, fruit cracking and internal breakdown are some important biotic and abiotic stresses associated with pomegranate cultivation. Pomegranate trees are attacked by about 45 species of insects and fruit is most vulnerable to the attack of pest. At present, we still lack development of eco-friendly pest management practices using new molecules .The fruits are attacked by several physiological disorders and insects, which is the main cause for decline of its production. Aspects like pomegranate butterfly incidence and cracking are the limiting factors in pomegranate cultivation and are prevalent throughout the country. The control of pest by chemical means provides partial solution. Therefore, obvious choice is to maintain a natural balance for growing best variety with desired characters and resistance. There is need to fully exploit the potential of high yielding varieties with resistance to pomegranate butterfly, fruit borer and stem borer and other biotic and abiotic stresses.

Fruit cracking is mainly associated with fluctuation of soil moisture, day and night temperatures, relative temperatures, relative humidity and of rind pliability. The disorder is reported to be due to boron and calcium deficiency. There is further attack of insects or fungal attack on the cracked fruits. So cracked fruits become unfit for marketing. Prolonged drought generally causes hardening of the peel. If this is followed by irrigation or rains, the pulp grows fast and ultimately the peel cracks. Pant (1976) observed that air temperature rise was cause of fruit cracking. It may amount to 63 per cent in the spring crop (January-June), 34% in the winter crop (October-March) and only 9.5% in the rainy season crop (July-December). The fruit cracking was quite high in almost all the cultivars as reported by Malhotra *et al.,* (1983). Fruit drop is one of the main problems and plant growth regulators may be exploited to control the problem and also to improve the fruit weight and quality significantly.

Policy Issues

Farmers of Rajnandgaon, Kawardha and Durg districts of Chhattisgarh can successfully introduce pomegranate cultivation for domestic and export purpose through their associations with adjoining pomegranate producing areas of MP and Maharashtra. Government intervention in providing technology and training can give boost to the farmers' economy through pomegranate cultivation for export purposes as well. Of late, problems due to diseases such as bacterial blight and pomegranate wilt have had a deleterious effect on the crop in the producing areas. Intervention in providing subsidies to the farmers, creating infrastructure facilities for production including establishment of modern nursery and transport to primary markets, post-harvest management and marketing and standardization of packaging techniques and research support are the aspects which need special attention and which will go a long way in developing pomegranate crops in the dry areas of Chhattisgarh and extending help to the farmers.

16

Status of Pomegranate (*Punica granatum* L.) Cultivation in Himachal Pradesh

N. Sharma[1], Prativa Sahu[2] and R.C. Sharma[1]

[1]Dr YS Parmar University of Horticulture and Forestry, Solan-173 230, Himachal Pradesh
[2]ICAR-National Research Centre for Pomegranate, Solapur, 413 265, Maharashtra

Introduction

Pomegranate (*Punica granatum* L.) has become an important fruit crop of Himachal Pradesh in the last decade. Traditionally wild pomegranate, which resembles cultivated pomegranate for various morphological characters, has been one of the most important fruits of the state (Sharma and Sharma, 1990). In India, it is found in vast tract of the hill slopes of Himachal Pradesh, Jammu and Kashmir and Uttarakhand at an altitude of 900 to 1800 m above mean sea level. In Himachal Pradesh, it is found in some areas of Solan, Sirmour, Mandi, Shimla, Kullu and Chamba districts (Bhrot, 1998). It is mostly found in those areas where slightly hot climate characterized by dry summer and fairly pronounced winter prevails.

The existence of vast tracts of wild pomegranate in different parts of the state provides ample evidence for the scope of its successful cultivation. However, earlier attempts to popularize it did not yield desirable results, because of problems of *anar* butterfly, fruit splitting and inferior fruit quality. Recently, improvement in the production technologies and climate change has generated interest among farmers to adopt pomegranate as a commercial crop. Now it has become an important fruit crop in the sub-tropics to sub-temperate zone of Himachal Pradesh. In recent years, it has almost completely replaced apple crop in the Kullu valley as the later has failed and become unprofitable due to climate change. Summer in the lower valley areas of Kullu is warm and dry and is highly suitable for the production of high quality pomegranate. It is also cultivated in the districts of Mandi, Shimla, Hamirpur, Bilaspur, Sirmour, Solan and Kangra, however, fruit quality is inferior and productivity is low in comparison to Kullu district. Currently, it is cultivated in an area of about 1709 ha with over 749 MT annual production (Anonymous, 2013). The district Kullu alone accounts for about 18 % of area and 47 % of total production. As a result of success achieved from pomegranate cultivation, its production in the state is bound to increase further in future as farmers of lower areas are rapidly shifting from apple to pomegranate cultivation.

Varietal Status

Recommended varieties in Himachal Pradesh are Kandhari, Bedana, Chawla, Ganesh and Bhagwa. However, Kandhari is most adapted to the mid hill conditions of the state. It is high yielding variety, producing large, attractive pinkish red coloured and superior quality fruit; however, it is hard seeded. G-137 is suited for the mid hills and low valley areas of the state. Its fruit weighs between 225-275 g, has pinkish yellow to reddish yellow rind colour, having light pink arils, soft seeds and medium total soluble solids (TSS), but it is prone to cracking. Besides, discolouration of arils in ripened fruits is common in this cultivar. Late maturing 'Bhagwa' fruits are smaller in size weighing 250-300 g, very attractive, 'saffron' coloured; soft seeded and has higher TSS. Fruit splitting is somewhat low, however, has not adapted well under the sub- temperate climatic region of state. Apart from these, varieties like Achikdana, AnarShirin Mohammad Ali, Jalore Seedless, Jodhpur Red, KandhariHansi, Nabha, P-75-K-3, P-75-K-5, P-23, P-26 are under evaluation.

Besides, wild form locally known as 'Daru" has a great economic importance. The edible part of this fruit is a rich source of organic acids apart from having appreciable amount of sugars, anthocyanins, phenols, ascorbic acid, *etc.* Arils of this fruit also contain good amount of minerals like phosphorus, calcium, potassium and iron (Parmer and Kaushal, 1982). However, it is too acidic by nature, as such cannot be used for table purpose; mostly being used as a good souring agent in curries, chutneys and other culinary preparation. Its seeds are sun dried to make *anardana*. Besides *anardana,*huge quantities of the fruit rind is exported for utilization in various industries. One of the important centers of wild pomegranate in HP is Darladghat, located about 50 km from Shimla and its name literally means a pass or place of *daru* (wild pomegranate) trees. A village "Darwa" in district Sirmour denotes growing of *daru* in the surroundings.

The *Daru* was employed in breeding for anardana variety and disease resistance as it exhibited high level of acidity and field tolerance for bacterial nodal blight caused by *Xanthomonas axonopodis* pv. *punicae*. In order to achieve these objectives, Daru was crossed with Ganesh (C1) and F1 of Ganesh *Nana (C2) and Ganesh *Kabul Yellow (C3). Most of the hybrids are close to *Daru* with respect to plant morphology.

Propagation

Pomegranates can be propagated from hardwood or semi-hard wood cuttings. In case of raising of plants by hard wood cuttings, the cuttings of about 1.0 cm diameter having 15 – 20 cm length are taken in the month of February when the plants are dormant. For better rooting, lower end of cuttings are immersed in a solution of IBA @ 5000 ppm for 5-10 seconds before placing in the rooting media. Plants can also be raised by semi-hardwood cuttings of 0.75 – 1.0 cm diameter having length of about 15-20 cm taken during the second week of August and rooting is facilitated in the mist chamber. However, hard wood cutting is preferred means of propagation, as it is easy, less expensive and gives better rooting percentage and field survival rate.

Planting

One-year-old rooted cuttings are planted in the field in December-January at a spacing of 5m × 5m. Rooted plants along with earth ball can also be planted in rainy season, however, growth and survival percentage of plants are less. At the time of planting, 10 kg of well rotten farm yard manure and 500 g single super phosphate are mixed with soil of each pit. For the protection against collar rot, plants are treated with copper fungicide prior to planting. Irrigation is given immediately after planting.

Training and Pruning

Multi-trunk training system by developing strong 4-6 main stem arising from the ground is followed. These stems are kept bare up to 20 cm from the soil surface by periodically removing any emerging shoots. Suckers are removed as and when noticed. Annual pruning is done in December-January, which involves removal of crowded and diseased shoots.

Manure and Fertilizer Application

One-year old plant is given 10 kg FYM, 125 g N, 100 g P_2O_5 and 150 g K_2O. These doses are increased every year until plants are 3-5 year old and thereafter these are stabilized. Recommended dose for fully grown plant is 50 kg FYM, 625 g N, 240 g P_2O_5 and 600 g K_2O. Full dose of FYM, phosphorus and potash and half dose of nitrogen are applied in January. The remaining half dose of N is applied in two split doses, at the time of fruit set and 4-5 week after fruit set.

Irrigation

Though, pomegranate is tolerant to water stress conditions, it responds well to irrigation. Irrigation at regular interval during summer not only increases fruit size and yield, but also minimizes the incidence of fruit splitting. In Kullu valley, basin irrigation is practiced. In rain fed orchards, *in situ* moisture conservation with 10 cm thick hay mulching improves yield and fruit quality (Sahu *et al.*, 2013).

Production Constrains

Fruit Cracking

Fruit cracking is a serious problem in pomegranate, causing economic losses to the extent of 60-70 per cent to the growers. It may be due to boron deficiency in young fruits, while in case of developed fruits; long dry spell followed by rains is the major cause of splitting. However, it may also occur due to fluctuation in day and night temperature. In cultivar G 137, foliar application of 0.2% boron or 10 ppm forchlorfenuron (CPPU) in mid-May and mid-June minimized the incidence of fruit cracking under the mid-hill conditions of Himachal Pradesh (Sharma and Belsare, 2011). In the rain fed conditions of the mid-hills, soil working technique "crescent bund with open catchment's pit" + foliar application of 5 ppm CPPU or 0.2% H_3BO_3 in mid-May and mid-June significantly decreased the fruit cracking in pomegranate cultivar Kandhari (Sahu *et al.*, 2013).

Insect Pest

In Himachal Pradesh, anar butter fly (*Deudorix isocrates*) is the major pest of pomegranate. Insect first lays eggs on flower or newly set fruit, subsequently caterpillars bore inside the developing fruits and feed on pulp and seeds. Later on damaged fruit is infected by bacteria or fungi, causing fruit rot. Application of cypermethrin @ 0.01% or monocrotophos @ 0.036% in second week of June and in first and third week of July is recommended for its control.

Diseases

Pomegranate orchards are adversely affected by various foliar (leaf spots), fruit (spots and rots) and soil borne (wilt) diseases. In Kullu valley, the biggest threat to pomegranate cultivation is posed by wilt disease caused by *Ceratocystis fimbriata* and *Fusarium oxysporum* with their incidence varying from 1.03 to 15.3 and 0.1 to 7.3 percent, respectively (Khosla and Bhardwaj, 2014). The problems were more severe in areas having heavy soils. Incidence of leaf and fruit spots caused mainly by *Cercospora punicae* (*Pseudocercospora punicae*) and *Alternaria alternate* is common. Similarly, fruit rot was caused by *Coniella granati* (1.0 to 14.8%), *Phomopsis aucubicola* (1.0 to 14.7%) and *Phytophthora sp.* (1.4 to 13.6%). Bacterial pathogen *Xanthomonas axonopodis* pv. *punicae* has been found causing leaf and fruit spots and subsequent fruit rotting and stem and twig canker at different locations (Hathithan in Kullu district and Shilligad of Mandi district) on Mridula and Bhagwa varieties.

Harvesting and Packing

The fruits are harvested when skin develop proper colour characteristic of respective cultivars. Harvesting is done manually, two or four times at weekly interval when fruits from different flushes mature properly. Harvesting begins with during the end of August and continue through to October or early November depending on locations and cultivars. After harvesting, the fruits are graded according to the size and are packed in 10 kg capacity cardboard boxes for transportation and marketing.

17

Paradigm of Healthy Pomegranate Orchards Establishment in North Karnataka

R.M. Hosamani, A.R. Sataraddi, P.S. Pattar,
R. Veeranna, Shivakumar Ekbote, K.B. Yadahalli,
S.C. Angadi and M.A. Gaddanakeri

Krishi Vignyan Kendra, Bagalkot-587 101, University of Agricultural Sciences,
Dharwad, Karnataka
Email: rmhosamani@gmail.com

Introduction

Pomegranate (*Punica granatum* L.) is a delicious fruit with great preference from consumers. It is an important fruit crop both from economic point of view to farmers, industry and governments at centre and state besides, nutritional and medicinal properties contribute to well being of nation's population. In India, this crop is grown in many states. Maharastra is the largest grower with nearly 75 % of the area and production. Karanataka is the second largest pomegranate growing state in terms of area and production.

Koppal, Bagalkot , Bijapur , Bellary districts followed by Gulburga, Raichur, Gadag, Belgavi, *etc.* in North Karnataka though it's also grown in Southern Karnataka districts like Chitradurga, Tumkur, Kolar, *etc.* The quality of fruits grown here are sought by major markets in the country and abroad (Hosamani, 2011).

An analysis of the trends, problems, solutions, technologies, their impact on pomegranate cultivation has been attempted in this chapter.

Materials and Methods

Area and production of the pomegranate has been collected and analyzed to know the trends and the reasons for the same identified. Survey of farmers has been done and SWOT analysis done. Problems in pomegranate cultivation have to be identified. Technologies pertinent to tackle the problems have been assessed for refinement and front line demonstrations were taken up in target areas. Results analysed and economics of the benefits worked out.

Results and Discussion

Area under pomegranate in Bagalkot district was 1686 ha in 2006-07. It increased to 2049 ha in 2008-09 but declined to 1931 ha in 2009-10 and then further declined to 1687 ha in 2010-11. The fruit production has increased from 15273 tons (2006-07) to 23859 tons (2009-10) but declined to 15692 tons in 2010-11. The -productivity has increased from 9.05 t/ha (2006-07) to 12.36 t/ha (2009-10) but declined to 9.3 t/ha in 2010-11.

Rapid increase in pomegranate area has been witnessed due to its higher income and profitability in last few years though higher investment is needed in terms of money, time and labour from farmer's side. These areas saw a rapid decline in area of pomegranate cultivation on account of severe bacterial blight followed by wilt and shot hole borer infestation. Bacterial blight disease caused by *Xanthomonas axonopodis* pv. *punicae* is widely prevalent occurring in a devastating proportion causing fruit losses to the extent of 70-100 percent (Anandgoud *et. al.*, 2014) wiping out orchards making farmers go bankrupt.

Bacterial blight out break was due to recklessness in the plant multiplication process and hurried manner of planting in field ignoring caution and procedure inlaid in the recommended package of practices.

This year (2013-14) there has been higher rainfall with longer duration cloudy weather after lapse of nearly four to five years creating congenial conditions for outbreak of bacterial blight disease in traditional and non-traditional growing areas. This has led to increase in bacterial blight incidence and rapid progress of infected orchards even under regular management practices by some of the progressive farmers. Many of the newly established orchards had nursery carried inbuilt inoculums exploding like time bomb with congenial conditions leading to many of the farmers uprooting their plantations.

Krishi Vigyan Kendra, Bagalkot, Bijapur as well as University of Agricultural Sciences, Dharwad, Raichur and University of Horticultural Sciences, Bagalkot have been in forefront in taking pertinent technologies to tackle the farmers problems.

Krishi Vigyan Kendra, Baglakot was established in 2005-06. Since then scientists have been involved in promotion of pomegranate cultivation among enthusiastic progressive fruit growing farmers Congenial dry climatic conditions beneficial to quality fruit production.

Poor quality and diseased planting material, bacterial blight disease, anthracnose disease, shot hole borer, wilt, fruit cracking, unscientific cultural practices, improper nutrient management, *etc.* were identified as limiting the productivity. Non-availability of disease free planting material was a major factor in bacterial blight and anthracnose spread.

Major problems identified by Krishi Vigyan Kendra, Bagalkot in the practices of pomegranate growers are i) Planting material is not disease free in most cases, ii) Planting is not done in recommended sized pits, iii) Pits are not filled with adequate amount of FYM, sand silt in most cases, iv) Recommended bioagents and neem cake is not used in required quantity, v) Plant protection measures are not need based, vi) Balanced nutrient application is missing, *etc.*

KVK has been taking technologies to farmers to improve their production, productivity and income through frontline demonstrations, on campus training, off campus training programmes, campaigns, field days, folders, pamphlets, radio talks, exhibitions, *etc.* To tackle the shortcomings, bottlenecks , weaknesses, and to boost up productivity SWOT analysis was done. KVK, Bagalkot formulated front line demonstration (FLD) and on farm testing (OFT) demonstrations, on campus and off campus training programmes, distribution of relevant cultivation information through printing of folders, booklets and use of mass media.

Some of the frontline demonstrations completed are on i) Use of boron and micronutrients to control fruit cracking and for quanlity fruit production, ii) Integrated disease management to control bacterial blight, iii) IPM for fruit borer.

Fruit cracking in pomegranate was tackled through OFT's (Table 17.1). Use of 0.2% boron spray at flowering and spray of 0.2% boron plus 0.2% $ZnSO_4$ at fruit development increased marketable uncracked quality fruits by 26% over farmers practice. There by consumers were benefited with greater proportion of attractive edible fruits and farmers net returns were Rs. 5,43,637/ha with BC ratio of 1:12.25.

Table 17.1: Pomegranate Fruit Cracking Management OFT Assessment in Bagalkot District During 2008-09.

Production Technology assessed	Unit (t/ha)	Net Return (Profit) in Rs. / ha	BC Ratio
Technology option 1 (Farmer's practice): Irregular spray of micronutrients	7.8	4,31,460	1 : 12.80
Technology option 2: Spraying of Boron (0.1%)	8.0	4,38,542	1 : 10.56
Technology option 3: a) Spraying of 0.2% Boron at flowering stage, b) Spraying of 0.2% Boron along with zinc 0.2% at fruit development stage	9.8	5,43,637	1 : 12.25

Bacterial blight disease incidence has been very devastating on pomegranate in Bagalkot district as well as at national level. Rapid area increase using seedlings from diseased orchards greatly contributed to its spread, besides improper management practices caused elimination of many orchards on account of drastic reduction in yields. Frontline demonstrations on integrated disease management of bacterial blight was undertaken and was effective (Table 17.2 and Table 17.3). Yields increased by 27 per cent resulting in profit of Rs. 4.05 lakhs per ha and a BC ratio of 1:10.0 compared to farmers practice fetching Rs. 1.40 lakhs per ha on a average with a BC ratio of 1:4.50.

Table 17.2: Pomegranate Productivity Boosting through Integrated Management of Bacterial Blight Front Line Demonstrations.

Name of the technology demonstrated	Variety	Year	Farming situation	No. of Demo.	Area (ha)	Yield (q/ha) Demo. H	L	A	Check	% Increase
IDM for bacterial blight	Kesar / Bhagwa	2007-08	Irrigated	5	2	163.0	87.0	150.0	60.0	15.00
IDM for bacterial blight	Kesar	2008-09	Irrigated	5	2	42.5	31.5	35.0	27.5	27
IDM for Anthracnose	Kesar	2008-09	Irrigated	5	2	37.5	30.0	32.8	25.0	31.0

Table 17.3: Economics of Pomegranate productivity boosting through Integrated management of Bacterial blight front line demonstrations.

Name of the technology demonstrated	Year	*Economics of demonstration (Rs./ha)				*Economics of check (Rs./ha)			
		Gross Cost	Gross Return	Net Return	BCR	Gross Cost	Gross Return	Net Return	BCR
IDM for bacterial blight	2007- 08	45000	450000	405000	10.0	40000	180000	140000	4.5
IDM for bacterial blight	2008-09	30800	100500	69700	3.26	28000	82000	54000	2.9
IDM for bacterial blight	2009- 10	29300	98400	69100	3.33	27400	75000	47600	2.73

To tackle the challenge from anthracnose disease, KVK, Bagalkot formulated front line demonstrations on integrated disease management. This resulted in increased quality fruits yield upto 51.8 per cent (Table 17.4). bringing in net profit of Rs. 1.06 lakh per ha with a BC ratio of 1:3.54 compared to farmers practice making net profit of Rs. 0.63 lakh per ha with a BC ratio of 1:2.80 (Table 17.5).

Table 17.4: Pomegranate Productivity Boosting Through Integrated Management Of Anthracnose Szdisease Front Line Demonstrations

Name of the technology demonstrated	Variety	Year	Farming situation	No. of Demo.	Area (ha)	Yield (q/ha)				% Increase
						Demo			Check	
						H	L	A		
IDM for Anthracnose	Kesar / Bhagwa	2009- 10	Irrigated	5	2	37.5	30.0	32.8	25.0	31.0
IDM for Anthracnose	Kesar	2010- 11	Irrigated	5	2.0	5.5	3.25	4.25	2.8	51.8

Table 17.5: Economics of Pomegranate Productivity Boosting through Integrated Management of Anthracnose Disease Front Line Demonstrations

Name of the technology demonstrated	Year	*Economics of demonstration (Rs./ha)				*Economics of check (Rs./ha)			
		Gross Cost	Gross Return	Net Return	BC Ratio	Gross Cost	Gross Return	Net Return	BC Ratio
IDM for Anthracnose	2009-10	29300	98400	69100	3.33	27400	75000	47600	2.73
IDM for Anthracnose	2010-11	42000	148750	106750	3.54	35000	98000	63000	2.8

To tackle the damage from thrips, KVK, Bagalkot formulated on farm testing on integrated pest management involving spray of Fipronil 5 EC @ 2 ml/liter. This resulted in increased quality fruits yield upto 27 per cent bringing with a BC ratio of 1:4.6 compared to farmers practice making net profit of Rs. 0.63 lakh per ha with a BC ratio of 1:3.26 (Table 17.6).

Table 17.6: Pomegranate Thrips Management OFT Assessment In Bagalkot District During 2010-11.

Technology Assessed	Production (t/ha)	Net Return (Profit) in Rs. / ha	BC Ratio
Technology option 1 : Farmer's practice	11	5,25,000/-	1:3.62
Technology option 2 : Spray of Imidacloprid @ 0.3 ml/lit.	15	7,25,000/-	1:4.12
Technology option 3 : Spray with Fipronil 5 SC @ 2 ml/lit	17	9,00,000/-	1:4.6

These efforts of KVK, Bagalkot has succeeded in creating awareness of importance of adopting integrated crop management practices and integrated disease management practices in pomegranate that is reflected in increased area and sustained productivity despite increased major diseases pressure on the crop.

Presently, FLD's are focusing on use of disease free planting material i.e. use of tissue culture seedlings with recommended pit sizes (2'x2'x2'), use of bioagents (25 g of *Trichoderma, Pseudomonas,* PSB, *Azospirillum, Azatobacter, Mycorrhiza,*), neem (0.5 kg), vermicompost (1 kg), FYM (10 kg), pit dusting with boric powder (200 g) per pit along with ICM and IPM are immediate starting measures with long term impact to minimize bacterial blight and wilt ultimately leading to healthy orchards and happy farmer .

18

Production of Pomegranate by Marginal Farmers as Sustainable Entrepreneurship

S.N. Ambad and P.S. Chandane

Agricultural Technical School, Solapur - 413 002, Maharashtra
E-mail: snambad@mail.com

Introduction

Pomegranate (*Punica granatum*) is one of the commercially important fruit crops of India. Maharashtra is leading in production of pomegranate. The area under crop in Maharashtra is 128650 hectares with production of 1197710 MT. In Maharashtra it is grown in the districts of Solapur, Ahmednagar, Nashik, Pune, Beed, Osmanabad, Jalna, Aurangabad and Jalgaon. The crop has capacity to grow in wide range of climatic conditions. It is grown both in dry and irrigated areas. Other pomegranate growing states are Karnataka, Andhra Pradesh, Gujrath, Rajasthan, Uttar Pradesh and Haryana. It is also exported to the gulf countries *viz.* Bahrain, Kuwait, Oman, Saudi Arabia, UAE and Netherlands. The fruit has medicinal properties and also used for preparing many processed products. The farmers of Solapur district in drought prone area are undertaking the professional cultivation of pomegranate as enterprise. It became the most remunerative crop for them.

Materials and Methods

The sample survey of the marginal and other farmers of the drought prone area of the Solapur was undertaken to study the entrepreneurial cultivation of pomegranate. Average rainfall of district is less than 750mm and is most erratic in nature. There is wide fluctuation in minimum and maximum temperature and it goes up to 44 ^{0}C during summer while minimum 10-12 ^{0}C. It persists in all types of soils. The pomegranate growing pocket is having light to medium soils with scarce sources of irrigation. The farmers in scarcity zone are poor and having low land holding. Their economic resources are meager and deprived of good income. In these circumstances farmers of this dry part strive hard for their livelihood with traditional crops like jowar, bajra, sunflower, groundnut and vegetables. They have now diverted to the sustainable cultivation of pomegranate. It is not only the high income group but also the poor marginal farmers, unemployed youths, entrepreneurs, graduates and non graduates, skilled and non skilled persons have adapted pomegranate cultivation as most profitable farming enterprise. Even though there is acute shortage of water in the scarcity zone, the farmers with their hardship and intensive cultivation of pomegranate turned in to precision mode of commercial cultivation. Earlier grape was supposed to be the premier crop for economic returns but it was confined to handful of rich and progressive farmers. The case study with samples from scarcity zone was conducted to reveal the success of cultivation of pomegranate as enterprise.

Results and Discussion

The study indicated that poor marginal farmers, unemployed youths, entrepreneurs, graduates and non graduates, skilled and non skilled persons have adapted pomegranate cultivation as most profitable farming enterprise. They are following the intensive cultivation of pomegranate.

Soils: Shallow, medium and deep black soils, pomegranate predominantly planted in shallow and medium soils and deep black soils are used for sugarcane and banana.

Climate: The tropical climate with high temperature during summer and October. The plants remain evergreen and needs bahar treatment.

Varieties: The promising varieties are Bhagwa and Ganesh

Planting: plants raised from air layering are used for planting.it is done in July-August and February –March at the spacing of 4.5x3.0.Pits of 60cm^3 are dug a month prior to planting.

Irrigation and fertigation: Drip is most popular method adapted for irrigation. Recommended dose of fertilizers are applied through drip.

Bahar Treatment : In tropical climate plants remain evergreen and it needs regulation of flowering. It is done by conditioning plants to water stress or by leaf fall. In general either *Ambe bahar* or *Hasta bahar* is followed by the farmers. Fruits of *Ambe bahar* are better in quality. Depending upon the availability of water choice of bahar is done.

Training: Multistem system with three to four stem is maintained. Balanced and well distributed canopy necessary to hasten growth and fruiting.

Pruning: Removal of basal suckers, water shoots, cross branches and dead canes is done. Thinning and pruning of old spurs is done to encourage new growth.

Intercrop: During early stage of planting when plants are small short duration crops and vegetables are grown as intercrop which fetch about Rs.75000/-per hectare.

Diseases and Pests: Bacterial blight is the most important roblem farmers are facing these days and has became major hurdle in the production of pomegranate. Other main disease is leaf spot.The important pests are stem borer and shoot hole borer, fruit borer, mealy bug, aphid, white fly, and fruit sucking moth. The intensive pest management is adapted for control of the same. The pest and diseases can be solved with sanitation, precautionary measures and package of practices standardized by NRC and Agricultural University.

Harvesting: Fruits are harvested when the colour is changed to deep red or pink and attained full growth. Storage, grading is done as per shape and size. Fruits are stored at 5-10⁰C.

Sl. No.	Particulars	Rate (Rs)	Amount (Rs.)
1.	Yield/ha	20.31MT@Rs.60250 per MT	12,21,287/-
2.	Expenditure/ha	---	1,50,000/-
3.	Net profit/ha	---	10,73,677/-

Conclusion

Pomegranate has proved to be the most profitable crop of the farmers of Solapur district. Majority of the farmers of this dry part strive hard for their livelihood with traditional crops like jowar, bajra, sunflower, groundnut and vegetables have now diverted to the sustainable cultivation of pomegranate. Earlier grape was supposed to be the premier crop for economic returns but it was confined to handful of rich and progressive farmers. On the contrary pomegranate with an average production of 20.31 tonns/ha and a gross profit of Rs.12.21 lakhs with average rate of Rs.60 thousand per tone and production cost of Rs.1.50 lakhs is emerged as the first choice fruit crop of the farmers. It surpassed other fruit crops in net returns, export and also improved economic status and employment of the farmers of the scarcity district of Solapur. The farmers are well aware of the production technology of crop. Government agencies, agricultural universities, NRC on Pomegranate has developed suitable technologies for crop production. Besides this farmers are innovative and tech savvy. Pomegranate can be grown in wide range of soils .The production starts from very second year if managed well or otherwise from third year onwards and continues till 10-15 years. The menace of bacterial blight is the concern for the pomegranate growers; however it can be solved with sanitation, precautionary measures and package of practices standardized by NRC on Pomegranate and Agricultural University.

19

Identification of Nutrient Antagonism and Deficiency Symptoms in Pomegranate

R.A. Marathe and Y.R. Shinde

ICAR-National Research Centre on Pomegranate, Solapur 413 255, Maharashtra

Introduction

Pomegranate is very important fruit crop and assumes great significance under arid and semi-arid tracks of India. The cultivation of pomegranate is spread in an area of 106.6 thousand ha in India. Most of the pomegranate cultivation in this region is on very shallow, light textured and rocky soils having low moisture, nutrient holding capacity and fertility status of the soil (Marathe *et al.* 2006). In such soils, balanced and judicious use of fertilizer assumes great importance. Various methods as visual deficiency symptoms, soil nutrients analysis, tissue analysis has been used to diagnose nutrient requirement of the plants. Deficiency of any particular nutrient element is caused due to lowering of its content in the plant and t produced characteristic morphological and spectral symptoms particularly on the leaves of the plants (Jones and Benton, 1998). Characterization of these nutrient deficiency symptoms on leaves could be very effectively used by a trained personnel in diagnosing nutrient disorders and provide an immediate evaluation of nutrient status in plant. This information can be very effectively used in the areas which do not have any facility for soil and leaf nutrient testing. Till date there is no information on this aspect in pomegranate, hence present investigation was undertaken to characterize visual deficiency symptoms and synergistic and or antagonistic effect of different nutrients.

Materials and Methods

An experiment was conducted in sand culture under greenhouse conditions at the experimental research farm of ICAR-National Research Centre on Pomegranate, Solapur, Maharashtra. Well established, four month old, healthy pomegranate cv. Bhagwa seedlings having same growth and vigour were transplanted in plastic pots (8 kg capacity) filled with acid washed sterilized, inert white sand. The transplanted seedlings were grown initially with half strength nutrient solution of Hoagland formulation applying all 16 essential nutrients and increased to full length after fortnight supplied through irrigation water to meet evapotranspiration demand. The pots were flushed with double distilled water to remove accumulated salt in the sand. The nutrient treatments were arranged in completely randomised block design with five plants per treatment. The treatments were imposed using nutrient deficit technique comprises of -N, -P, -K, -Ca, -Mg, -S, -Fe, -Zn, -Cu, -Mn, -B and -Mo and one full nutrient strength solution. The plants were grown under 50% greenhouse irradiance conditions for six months and monitored periodically for vegetative performance and any deformities, physiological disorder or marked symptoms in leaves size and shape and documented in the form of photographs. At the end of experiment whole plants were uprooted, separated in roots, stem and leave washed thoroughly in sequence with water, liquid soap, acidic water and glass redistilled water and dried in shade for four days followed by oven drying at 70°C till constant weight. The dried samples were grounded, mixed well and used for analysing total N by microkjeldhal steam distillation method. The samples were digested in di-acid mixture (Chapman and Prat, 1961) and analysed for P using Vanadomolybdo phosphoric acid method, K flame photometrically, Ca and Mg by titrimetric method employing disodium salt of EDTA(Jackson, 1973) and micronutrients Fe, Zn, Mn and Cu using atomic absorption spectrophotometer (Perkin Elmer, USA make AAnalyst 400).

Results and Discussion

Plants grew normally and without any deficiency symptoms in the complete nutrient solution treatment indicating that the nutrient supply was proper and sufficient during the experimental period. Plant development and visual deficiency symptoms in the experimental treatments were as follows.

Macronutrient Deficiency Symptoms

In case of nitrogen deficiency general appearance of the plant was pale yellow with restricted growth as compared to plants supplied with all nutrients. Deficiency symptoms first appeared on lower and mature leaves. Young and new leaves became uniformly chlorotic. Yellowing appeared uniformly on whole leaf. Leaves became stiffer in strength, breakein to pieces on folding. Almost all the plants under this treatment produced large number of flower- buds and flowers (27 flowers / plant) at early stage and majority of them were hermaphrodite flowers.

Phosphorus deficiency symptoms first appeared on younger leaves. Yellowing of leaf starts from tip only other part remains green while in case of nitrogen deficiency, whole leaf becomes pale in colour. Leaves become slender, elongated and smaller in size. Leaf margins turns upwards and gets tunnel like shape. Plant growth gets stunted.

Potassium deficiency symptoms first appear on older leaves. Many brown spots appear on dorsal side of leaves along the leaf margin starting from tip. In advanced stage, leaf margin becomes yellow followed by scorching appearance.

Calcium deficiency symptoms appeared on younger leaves. Interveinal yellowing starts from leaf tip, advanced from margin towards midrib, veins remains green during initial stages, later stages it also become yellow. The yellow portion of the leaf tip proceeds in inverted 'V' shape. Pinkish tinge appears on lower yellow portion of the leaf. In advanced stage, yellow portion of leaves turn dark brown in colour and half of the leaf from the tip dry up.Thecolour of dry leaves is dark brown.

In case of magnesium deficiency leaf margin and vein appear light green in colour. Initially, grey patches appear on the side margin of the leaves and subsequently covers whole leaf. Leaves show drying up sign. The colour of dry leaves is grey.

Micronutrient Deficiency Symptoms

The experimental results showed that pomegranate is very hardy fruit crop and even after 3 to 4 four month very specific and prominent systems of nutrient deficiency were not observed on the saplings. After 5 to 6 months some deficiency symptoms were observed. In copper deficiency, margin of the leaf becomes yellow. Causes irregular growth and pale green leaves that wither at leaf margins. Leaves at top of the plant wilt first, followed by chlorotic and necrotic areas on leaves, and necrosis of the apical meristem. Leaves on top half of plant show unusual puckering with veinalchlorosis. In case of iron, initially deficiencies show as interveinalchlorosis in young leaves. Leaf veins remains green in colour, interveinalchlorotic mottling of immature leaves. In this experiment, no typical symptoms of iron deficiency were observed even after six month of the experimentation. But typical iron deficiency types of symptoms were observed in some other plants grown in pots and in field. Zinc deficiency first shows itself as pronounced interveinalchlorosis in young leaves and mid-shoot leaves. Interveinal yellowish areas started at leaf tip and margins and eventually affected all growing points of the plant. Greyish brown spots that form on leaves halfway up the plant and then spread. In case of zinc also deficiency symptoms were not much pronounced in pomegranate. Molybdenum deficiency showed pale leaves, with marginal leaf chlorosis. It can reveal itself as interveinal yellow spotting and mottling of older leaves.

Antagonistic/Synergistic Effects of the Nutrients

The data on leaf nutrient content (Table 19.1) revealed that deficiency of N, P, K, Fe and Cu nutrients recorded significant influence on the plant resulted in to lowest content of these nutrients in the leaves of pomegranate plant in the respective treatments. But such a strong impact was not observed in case of Ca, Mg, Mn and Zn nutrients. Deficiency of P adversely affected the uptake of majority of nutrients as N, K, Ca, Mg and Mn; Mn adversely affected the uptake of P; Cu adversely affected Mg and Fe content while B adversely affected Cu and Fe content in the leaves of pomegranate.

Similarly nutrient content in the stem of the pomegranate plants (Table 19.2) revealed that deficiency of N, P and K resulted into lower content of these nutrients in the stem of the plants while deficiency of Ca, Mg, Fe, Mn, Cu and Zn does not have much influence and maintained sufficient quantity of these nutrients in the stem in respective treatments. Deficiency of boron adversely affected the uptake of Mg, Fe, Mn, Cu and Zn in the stem of pomegranate plant. Similarly deficiency of P adversely affected the uptake of Ca; Mg adversely affected K uptake; S adversely affected Mn and Zn uptake; Cu adversely affected Mg uptake while deficiency of Zn adversely affected Mn uptake in the stem of pomegranate plant.

Table 19.1 : Leaf Nutrient Content as Affected by Deficiencies of Different Nutrients

Nutrient content →	N	P	K	Ca	Mg	Fe	Mn	Cu	Zn
Deficient nutrient ↓			(%)				(ppm)		
- N	2.34^b	0.140^{abc}	3.80^f	2.74^{ab}	0.68^{abc}	290.9^{bc}	216.4^{ab}	22.2^a	28.1
- P	3.35^a	0.037^f	3.08^g	1.99^f	0.54^{cd}	248.4^{bcd}	119.6^d	17.4^b	29.0
- K	3.74^a	0.104^{cde}	1.31^h	2.61^{abc}	0.75^a	227.7^{cd}	217.7^{ab}	16.8^b	27.1
- Ca	3.38^a	0.119^{abcd}	4.49^{cd}	1.99^f	0.46^d	297.3^b	223.0^{ab}	14.6^{bc}	26.6
- Mg	3.54^a	0.132^{abc}	5.02^{abc}	2.10^{ef}	0.55^{cd}	230.5^{cd}	191.9^{abc}	12.5^{cd}	26.8
- S	3.50^a	0.109^{bcde}	4.40^{de}	2.46^{abcd}	0.64^{abc}	365.5^a	198.2^{ab}	13.6^c	24.2
- Fe	3.43^a	0.147^{ab}	5.13^{ab}	2.79^a	0.72^{ab}	144.0^f	190.6^{bc}	8.5^e	30.8
- Mn	3.59^a	0.092^{de}	4.61^{bcd}	2.58^{abc}	0.60^{bcd}	236.5^{bcd}	199.3^{ab}	9.0^e	27.6
- Cu	3.51^a	0.129^{abcd}	4.84^{abcd}	2.42^{bcde}	0.55^{cd}	151.9^{ef}	194.0^{ab}	8.2^e	24.5
- Zn	3.69^a	0.149^a	5.25^a	2.63^{abc}	0.66^{abc}	371.4^a	195.3^{ab}	10.4^{de}	25.2
- B	3.64^a	0.123^{abcd}	4.60^{bcd}	2.68^{ab}	0.63^{abc}	212.8^{de}	227.9^a	8.3^e	29.9
- Mo	3.73^a	0.079^e	3.84^{ef}	2.34^{cde}	0.58^{bcd}	233.8^{bcd}	192.1^{abc}	8.7^e	31.2
ALL	3.58^a	0.139^{abc}	4.78^{abcd}	2.23^{def}	0.61^{abc}	290.4^{bc}	156.0^c	9.2^e	26.8
CD (p=0.05)	0.52*	0.038*	0.58*	0.335*	0.145	66.7*	36.1*	2.97*	NS

Table 19.2: Stem Nutrient Content as Affected by Deficiencies of Different Nutrients

Nutrient content→ / Deficient nutrient ↓	N	P	K	Ca	Mg	Fe	Mn	Cu	Zn
	(%)					(ppm)			
- N	1.45^g	0.325^{cd}	2.00^a	2.05	0.53	245.7^{de}	170.6^{ab}	39.8^a	51.4^a
- P	2.00^f	0.040^f	1.13^{efg}	1.94	0.43	323.4^b	182.4^{ab}	39.8^a	53.0^a
- K	2.68^{cd}	0.327^{cd}	0.45^h	1.67	0.46	268.0^{cd}	177.8^{ab}	31.2^{bcd}	52.4^a
- Ca	2.89^{bc}	0.345^{bcd}	0.99^{fg}	1.96	0.48	303.4^{bc}	160.1^{bc}	36.3^{ab}	53.9^a
- Mg	2.34^e	0.307^{de}	0.90^g	2.28	0.51	397.3^a	196.0^a	31.7^{bcd}	40.4^c
- S	2.69^{cd}	0.259^e	1.16^{defg}	1.75	0.45	206.8^{ef}	94.3^e	27.2^{de}	31.0^d
- Fe	2.60^d	0.527^a	1.43^{bc}	2.18	0.50	257.6^{cde}	176.4^{ab}	35.7^{abc}	50.5^a
- Mn	3.14^{ab}	0.385^{bc}	1.36^{bcde}	1.91	0.51	236.4^{def}	114.2^{de}	30.3^{cde}	41.5^{bc}
- Cu	2.83^{cd}	0.401^b	1.33^{bcde}	1.95	0.41	215.1^{def}	172.6^{ab}	28.7^{de}	41.3^{bc}
- Zn	3.16^a	0.377^{bc}	1.20^{cdef}	2.13	0.46	217.4^{def}	93.3^e	26.5^{de}	34.0^{cd}
- B	2.74^{cd}	0.336^{cd}	1.41^{bcd}	1.75	0.41	184.8^f	109.9^{de}	24.9^e	31.2^d
- Mo	3.09^{ab}	0.379^{bc}	1.37^{bcde}	1.80	0.45	241.3^{de}	134.1^{cd}	27.3^{de}	48.7^{ab}
ALL	2.27^e	0.310^{de}	1.53^b	1.86	0.46	266.3^{cd}	134.3^{cd}	29.9^{cde}	38.0^{cd}
CD (p=0.05)	0.26*	0.064*	0.26*	NS	NS	55.1*	33.7*	5.87*	7.81*

20

Eco-friendly Management of Bacterial Blight of Pomegranate Caused by *Xanthomonas axonopodis* pv. *punicae*

V.I. Benagi[1], Basamma[2], Madhu S. Giri[2] and V.B. Nargund[2]

[1]College of Agriculture, Hanumanamatti, Haveri-581 115, Karnataka
[2]Department of Plant Pathology, UAS, Dharwad-580 005, Karnataka
E-mail: hodenagi@gmail.com

Introduction

Pomegranate, *Punica granatum* L., is regarded as the "Fruit of Paradise" which can be found throughout the Mediterranean region, Southeast Asia, California and Arizona in USA. It is commonly known as *'Anar'* in Hindi and *'Dalimb'* in Kannada and it belong to the family Lytheraceae. In India, it is cultivated as a commercial fruit crop in Maharashtra, Karnataka, Andhra Pradesh, Gujarat, Madhya Pradesh and Uttar Pradesh (Chadha, 2001). However, this crop is grown all over India and as a component of kitchen and homestead garden. Area under pomegranate is increasing worldwide because of its hardy nature, wider adaptability, drought tolerance, higher yield levels with excellent keeping quality and remunerative prices in domestic and export markets. It thrives well in arid tropics and subarid tropics and comes up very well in soils of low fertility status as well as in saline soils.

Pomegranate is a good source of carbohydrates and minerals such as calcium, iron and sulphur. It is rich in vitamin-C and citric acid (Malhotra *et al.*, 1983). The fruits of pomegranate are known to posses pharmaceutical and therapeutic properties and are used as component of folk medical practices.

Cultivation of pomegranate in recent years has met with different traumas such as pest and diseases. Among the diseases infecting pomegranate, the bacterial blight caused by *Xanthomonas axonopodis* pv. *Punicae* (Hingorani and Singh) Vauterin *et al.* is as a major threat. It causes loss upto 70-80% (Benagi *et al.*, 2011). In Karnataka, survey report revealed that, 20-90 percent of disease severity in Bijapur and Bagalkot districts (Ravikumar *et al.*, 2006). Recent reports revealed the highest severity of tree was 74.80 percent in Bagalkot district and minimum severity of 6.73 percent in Bellary districts (Anon., 2008). During 2008-09 the disease has reached its alarming stage bringing substantial damage to the crop and heavy loss to the farmers.

Major component of present management practice of bacterial blight of pomegranate include use of bactericides. The continuous and indiscriminate use of these bactericides may result in development of super races of the pathogen which may develop resistance to currently available bactericides in the market. So, there is an immediate need for development of alternative strategy which avoids such tragedies in future. In recent years, there has been a major thrust on residue free organic pomegranate production and also to reduce the cost of cultivation of this crop. Plant nutrients also play an important role in susceptibility or resistant mechanism of the host to different pathogens. Taking the task into consideration, efficient bioagents, botanicals and micronutrients need to be explored to fit into the management schedule.

Materials and Methods

Isolation and Identification of *Xanthomonas axonopodis* pv. *punicae*

The *Xanthomonas axonopodis* pv. *Punicae* (*Xap*) was isolated and purified on Nutrient Glucose Agar (NGA) medium (Basamma, 2013). Further identification of the *Xap* was done using *gyrB* gene specific primer (Mondal and Kumar, 2011).

Isolation of Fluorescent Pseudomonads

The native fluorescent *Pseudomonads* were isolated from rhizosphere soils collected from pomegranate crop grown in Northern Karnataka and border areas of Maharashtra. Further, isolation was done according to the method of Weller and Cook (1983).

In vitro Screening of Fluorescent Pseudomonads against *X. axonopodis* pv. *punicae*

One hundred and twenty colonies of native fluorescent *Pseudomonads* were screened for their ability to inhibit the growth of *Xap* by following the dual culture assay (Ganesan and Gnanamanickyam, 1987).Observations were recorded for the production of zone of inhibition around the *Xap* and the diameter of the inhibition zone (DIZ) was measured after 72h. of incubation.

Elucidation of the Mechanism of Action of Native Fluorescent *Pseudomonads* against *X. axonopodis* pv. *punicae*

Detection of Fluorescin and Pyocyanin

Pseudomonas agar F favours the formation of fluorescein whereas *Pseudomonas* agar P stimulates the pyocyanin production and reduces fluorescein formation (King *et al.*, 1954).

Siderophore Production

Siderophore production by the fluorescent Pseudomonad strains was tested following the method of Schwyn and Neilands (1987).

HCN Production

Ability of the efficient fluorescent Pseudomonad strains to produce HCN was assessed as per the method of Wei *et al.,* (1991).

Antimicrobial Metabolite (Antibiotics) Production

The antibiotic activity of selected efficient fluorescent Pseudomonads strains was determined and assessed by extracting and testing the toxicity of metabolites produced by them following the method of Kraus and Loper (1992).

In vitro Evaluation of Plant Extracts

Plant based pesticides, which are relatively safe, economical and non-hazardous can be used successfully for the management of bacterial diseases in crop plants. The present investigation was aimed at screening seven plant extracts *viz.*, garlic, soapnut, meswak, adathoda, tulsi, nerium, neem for their antibacterial properties against *Xap*.

Fresh plant materials were collected and washed in distilled water, 100 grams of fresh sample was chopped and macerated in a sterilized pestle and mortar by adding 100 ml of distilled sterile water (1:1 w/v). The extract was filtered through Whatman's no. 1 filter paper; filtrate thus obtained was used as a stock solution. To study the antibacterial mechanism of plant extracts, inhibition zone assay method was followed.

A loop full (72 hours old) of *Xap*, multiplied on nutrient glucose agar (15ml) in petriplates was mixed with 5ml distilled sterilized water. 300µl of this solution was spread over nutrient glucose agar. Sterilized filter paper discs (Whatman's no. 1) measuring 5 mm diameter were soaked for 10 minutes in 5 percent and 10 percent of different plant extracts and placed on the surface of the nutrient glucose agar medium which was spread with *Xap* contained in the petriplates. The inoculated plates were incubated at 28 °C for 72 hours.

Observations were recorded by measuring the diameter of the inhibition zone around the filter paper disc in each plant extract at different concentrations in millimeter and data obtained was analysed statistically.

In vitro Evaluation of Bioagents on the Growth of the *Xanthomonas axonopodis* pv. *punicae*

The five *Pseudomonas fluorescens* strains and eleven PPFM strains were obtained respectively from Department of Molecular Biology & Biotechnology, and Department of Agricultural Microbiology, Agricultural College, Dharwad, University of Agricultural Sciences, Dharwad weretested for their inhibition effect on the growth of *Xap* by inhibition zone assay method as mentioned by above method.

Pseudomonas fluorescens was grown and maintained on King's B broth and PPFM strains (phylloplane bacteria) on AMS (Ammonium Mineral Salt) broth.

In vitro Evaluation of Micronutrients on the Growth of *Xanthomonas axonopodis* pv. *punicae*

Effect of micronutrients for their ability to inhibit the growth and development of *X. axonopodis* pv. *Punicae* and to control the bacterial blight of pomegranate were tested. Following are the nutrients and their concentrations which were tested *in vitro viz.*, Boron, Calcium, Iron, Magnesium, Micron special (boron, $MgSO_4$ and $ZnSO_4$), Zinc, Manganese, Sulphur, Copper at 0.5 percent concentration and inhibition zone assay was done by following the above mentioned method.

Field Evaluation of Selected Botanicals, Microbial Antagonist and Nutrient for Management of Bacterial Blight of Pomegranate (cv. Bhagwa)

Two field trials were conducted during *Mrig bahar* at Koppal district to study the combination effect of botanicals, bioagents, micro and secondary nutrients. Nine sprays were given at fifteen days interval and disease severity was recorded on fruits.

Results and Discussion

Isolation and Identification of *Xanthomonas axonopodis* pv. *punicae* through *gyrB* gene Specific Primer

The causal organism was isolated from the infected leaf by tissue maceration method using NGA medium. Isolation of the pathogen showed typical, yellow, mucoid bacterial colonies on NGA medium after 72 h. of incubation at 28 ± 1^0C. Culture of bacterium was purified by streaking single colony on to the NGA medium. The observations made pertaining to the isolation and pathogenicity in the present investigation were in conformity with the work of Manjula (2002), Yenjerappa (2009) and Basamma (2013).

The PCR assay performed for *gyrB* gene by using the primer set *gyr*BF-*gyr*BR produced DNA products of the predicted size i. e. 491bp or approximately 500bp which confirmed the identity of the pathogen as *Xanthomonas axonopodis* pv. *punicae*. The result is in accordance with the findings of Basamma (2013) who also got similar result when *gyrB* gene primer set was used for confirmation of all the 18 isolates of *X. axonopodis* pv. *punicae* collected from Karnataka, Maharashtra and Andhra Pradesh.

In vitro Screening of Fluorescent Pseudomonads against *X. axonopodis* pv. *punicae*

Among one hundred and twenty colonies of native fluorescent Pseudomonads tested for inhibition of *Xap, nine were found to be very effective antagonists.* The strain FP 128 produced the highest ZOI of 43.5 mm dia. followed by FP-85, FP-19, FP-03, FP-07, FP-117, FP-10, FP-135 and FP-14. Mishra and Arora (2011) evaluated the fifty four isolates of rhizospheric *Pseudomonas* and *Bacillus* as biocontrol tool for *Xanthomonas campestris* pv.*campestris.* Two isolates namely, KA19 and SE, with inhibition radius >11 mm were selected for further studies. Elucidation of the mechanism of action of native fluorescent Pseudomonads against *X. axonopodis* pv. *punicae*

Detection of Fluorescin and Pyocyanin

It was found that all the fluorescent Pseudomonad strains tested were known to produce fluorescein and pyacyanin in varied proportion after 48 hr. of incubation. The strains FP-03, FP-07, FP-10, FP-14 and FP-19 showed strong pyocyanin production compared to remaining strains. Strong fluorescein production was observed in FP-07 and FP-19.

Siderophore Production

All the nine fluorescent Pseudomonads strains produced siderophores with variable proportions. Result of siderophore production indicated that strains FP-117 and FP-128 showed maximum activity of siderophore production compared to others strains.

HCN Production

Among the nine strains tested only three strains i.e., FP-85, FP-117 and FP-128 were found to be positive for hydrogen cyanide production.

Antimicrobial Metabolite (Antibiotics) Production

To ascertain the antimicrobial metabolite production by the strains, metabolic products of each strain was extracted and separated by TLC plates.All the seven strains produced more than one metabolite, which appeared as blue, green and red spots under UV light (356 nm). The Rf values of the metabolites varied from 0.23 to 0.86. All the unidentified metabolites on TLC plates were scraped, eluted and again dissolved in 70 percent methanol and tested for *in vitro* inhibition of *Xap* by standard paper disc method.All extracted metabolite failed to show inhibition individually except a metabolite produced by strain FP-85 with Rf value of 0.52.

In vitro Evaluation of Plant Extracts on the Growth of *Xanthomonas axonopodis* pv. *punicae*

Among the seven plant extracts tested, soapnut was recorded highest inhibitory zone of 1.97cm at 10 percent and 0.95cm at 5 percent concentration which is significantly superior over other plant extracts. The extracts of garlic, miswak and adathoda inhibited the growth of *X. axonopodis* pv. *punicae* producing the inhibitory zone of 0.90cm, 0.80cm and 0.83cm at 10 percent and 0.72cm, 0.70cm and 0.62cm at 5 percent concentration, respectively. The extracts of neem, nerium and tulsi failed to inhibit the growth of bacterium even at 10 percent concentration. Among combination of effective botanicals, adathoda + soapnut recorded inhibitory zone of 1.47cm and 1.00cm at 10 and 5 percent, respectively followed by meswak + soapnut and adathoda + meswak (Table 20.1).

Table 20.1: Bio-efficacy of different plant extracts against *Xanthomonas axonopodis* pv. *punicae* under *in vitro* conditions

Plant extracts	Parts used	Inhibition zone (mean diameter in cm)		
		5%	10%	Mean
Garlic (*Allium sativum*)	Clove	0.73(1.31)*	0.90(1.38)	0.81(1.35)
Soapnut (*Sapindus mukorossi*)	Fruit	0.95(1.40)	1.98(1.72)	1.46(1.57)
Meswak (*Salvadora persica*)	Stem	0.70(1.26)	0.80(1.30)	0.65(1.28)
Adathoda (*Adathoda vessica*)	Leaf	0.63(1.27)	0.83(1.35)	0.73(1.31)
Tulsi (*Ocimum sanctum*)	Leaf	0.00(1.00)	0.00(1.00)	0.00(1.00)
Nerium (*Nerium oleander*)	Leaf	0.00(1.00)	0.00(1.00)	0.00(1.00)
Neem (*Azadirachta indica*)	Leaf	0.00(1.00)	0.00(1.00)	0.00(1.00)
Adathoda + Soapnut (*A. vessica + S. mukorossi*)	Leaf+ Fruit	1.00(1.41)	1.48(1.57)	1.24(1.50)
Adathoda + Meswak (*A. vessica + S. persica*)	Leaf+ Stem	0.68(1.29)	0.98(1.41)	0.83(1.35)
Meswak + Soapnut (*S. persica + S. mukorossi*)	Stem+Fruit	0.83(1.35)	1.18(1.47)	1.00(1.41)
Mean		0.49(1.22)	0.73(1.32)	0.61(1.27)
Source	SEM±	CD at 1%		
Plant extracts (P)	0.008	0.030		
Concentration (C)	0.004	0.014		
Interaction (P × C)	0.012	0.043		

* ×+1 transformed values

Manjula (2002) observed the significant difference in the inhibitory effect among the eight plant extracts screened against the growth of *X. axonopodis*pv. *punicae*. kalangi extract (1:1) was found more effective against the growth of Bengaluru fruit isolate followed by meswak, tulsi and patchouli, where as meswak exhibited highest inhibitory effect followed by kalangi and patchouli on the growth of Bijapur isolate. Later Jalaraddi (2006) found that both aqueous and alcoholic extracts of garlic, meswak and citronella were inhibitory to growth of the *Xanthomonas axonopodis* pv. *punicae*. Similarly, Yenjerappa (2009) reported that among the different plant extracts tested against *X. axonopodis* pv.*punicae*, garlic extract (10%) produced the maximum inhibition zone followed by parthenium, lantana leaf extract and onion bulb extract each at 10 percent concentration in *in vitro* condition.

In vitro Evaluation of Antagonistic Bacteria on the Growth of *Xanthomonas axonopodis* pv. *Punicae*

Among the five different strains of *Pseudomonas fluorescens* tested strain no. 326 (4 cm) (3.10 cm) and 139 (3.00 cm) has recorded highest inhibitory zone and these are significantly superior over other strains tested and there is no significant difference between these two strains. Followed by strain pf1 (2.77 cm), 134 (2.72 cm) and pf5 (2.65 cm) (Table 20.2).

Table 20.2: *In vitro* evaluation of *Pseudomonas fluorescens* against *Xanthomonas axonopodis* pv. *punicae.*

Pseudomonas fluorescens strains	Inhibition zone (mean diameter in cm)
134	2.72 (1.93)*
PF-5	2.65 (1.91)
326 (4)	3.10 (2.02)
PF-1	3.00 (2.00)
139	2.77 (1.94)
Mean	2.37 (1.80)
Sem±	0.03
CD	0.13

*Ö × +1 transformed values

The antagonism of *Pseudomonas fluorescens* against some *Xanthomonas* spp. was reported by Unnamalai and Gnanamanickam (1984), Sakthivel *et al.,* (1986), Safiyazov *et al.,* (1995). Jones *et al.,* (2011a) reported that, out of 62 fluorescent Pseudomonads evaluated, following the dual culture assay, 22 isolates exhibited antagonistic activity against *Xanthomonas axonopodis* pv. *Punicae* with inhibition zone varied from 2.0 to 5.0 mm.

Strain	Inhibition zone (mean diameter in cm)
20A	0.00 (1.00)*
71LIN	0.00 (1.00)
32	0.00 (1.00)
10L	0.95 (1.40)
75L	0.80 (1.34)
42U	0.00 (1.00)
23A	0.00 (1.00)

Contd...

Strain	Inhibition zone (mean diameter in cm)
80L	0.00 (1.00)
26U	0.00 (1.00)
38U	0.00 (1.00)
H5	0.00 (1.00)
CONTROL	0.00 (1.00)
Mean	0.15 (1.06)
SEM±	0.006153
CD at 1%	0.02

*Ö × +1 transformed values

Among the eleven different strains of PPFM tested, strain 10L has recorded highest inhibitory zone of 0.95 cm which is significantly superior over rest of the strains followed by 75L (0.80 cm). Other strains failed to show any inhibitory zone (Table 20.3).

Jones *et al.,* (2011b) reported that, Pink Pigmented Facultative Methylobacterium (PPFM) H5 strain, isolated from Bagalkot, has shown maximum inhibition zone (15.00 mm) followed by PPFM 42U (14.60 mm), PPFM 71 (14.40 mm) and PPFM 75L (14.00 mm) against *Xanthomonas axonopodis* pv. *Punicae* causing bacterial blight of pomegranate.

In vitro Evaluation of Micronutrients on the Growth of *Xanthomonas axonopodis* pv. *punicae*

Among the nine micronutrients tested, $CuSO_4$ has recorded highest mean inhibitory zone of 0.89 cm which is significantly superior over the other strains tested followed by micron special (boron, $MgSO_4$ and $ZnSO_4$) 0.41 cm at all concentrations tested. $MnSO_4$, $CaSO_4$, boron, sulfur, $FeSO_4$ failed to inhibit the pathogen. $CuSO_4$ has recorded maximum inhibition zone of 0.60 cm at 0.05 percent, 0.75 cm at 0.1 percent and 0.95 cm at 0.2 percent and 1.25 cm at 0.5 percent concentrations, respectively (Table 20.4).

Strong correlation between boron uptake in leaf tissues, residual boron in soil with black rot incidence in cauliflower was reported by Kumar and Kotur (1989). Exogenous supply of boron (0–6.4 mg/kg) in low boron containing altisols (hot water soluble soil containing boron of 0.1 mg/kg) indicated that, this micronutrient has a definite role in susceptibility of cauliflower to black rot caused by *Xanthomonas campestris* pv. *campestris*. Susceptibility was greatly observed in boron deficient (below 0.4 mg/kg) and boron excess (above 1.6 mg/kg) plants than the plants grown with optimum level of boron (0.4–1.6 mg/kg).

Dordas (2008) reviewed the effect of nutrients, such as N, K, P, Mn, Zn, B, Cl and Si, on disease resistance and tolerance. Among the micronutrients, Mn can control a number of diseases, as Mn has an important role in lignin biosynthesis, phenol biosynthesis, photosynthesis. Zn showed decreased or increased susceptibility to disease in some cases. B was found to reduce the severity of many diseases because it has function on cell wall structure, plant membranes and plant metabolism. Nutrients can reduce disease to an acceptable level, or at least to a level at which further controls by other cultural practices or conventional organic biocides are more successful and less expensive.

Field Evaluation of Selected Botanicals, Microbial Antagonist and Nutrient for Management of Bacterial Blight of Pomegranate (var. Bhagwa)

The percent fruit infection after 9th spray was minimum in T_5 (*Pseudomonas fluorescens*) and T_9, (copper oxychloride @0.3 % and streptocycline 0.05 %) whereas the highest percent disease incidence was observed in control plots (51.25%). Average fruit yield pooled over two locations indicated that significantly higher yield per tree was in T_9 and T_4 which were on par with each other (Table 20.5). The maximum C:B was observed in treatments T_9 and T_4 with value of 1:5.3 and 1:4.8 respectively (Basamma *et al.*, 2014).

Table 20.4: *In vitro* Evaluation of Micronutrients against
***Xanthomonas axonopodis* pv. *punicae*.**

Treatment	Inhibition zone (cm) Concentration (%)				
	0.05	0.1	0.25	0.5	Mean
	0.00	0.00	0.00	0.63	0.16
ZnSO₄	(1.00)*	(1.00)	(1.00)	(1.27)	(1.07)
CuSO₄	0.60	0.75	0.95	1.25	0.89
	(1.26)	(1.32)	(1.40)	(1.50)	(1.37)
MnSO₄	0.00	0.00	0.00	0.00	0.00
	(1.00)	(1.00)	(1.00)	(1.00)	(1.00)
MgSO₄	0.00	0.00	0.00	0.60	0.15
	(1.00)	(1.00)	(1.00)	(1.26)	(1.07)
CaSO₄	0.00	0.00	0.00	0.00	0.00
	(1.00)	(1.00)	(1.00)	(1.00)	(1.00)
BORON	0.00	0.00	0.00	0.00	0.00
	(1.00)	(1.00)	(1.00)	(1.00)	(1.00)
SULFUR	0.00	0.00	0.00	0.00	0.00
	(1.00)	(1.00)	(1.00)	(1.00)	(1.00)
FeSO₄	0.00	0.00	0.00	0.00	0.00
	(1.00)	(1.00)	(1.00)	(1.00)	(1.00)
MICRON SPECIAL	0.00	0.00	0.65	1.00	0.41
(boron, ZnSO4, FeSO4)	(1.00)	(1.00)	(1.28)	(1.41)	(1.17)
	0.00	0.00	0.00	0.00	0.00
Control	(1.00)	(1.00)	(1.00)	(1.00)	(1.00)
	0.06	0.1325	0.21	0.395	0.161
Mean	(1.03)	(1.05)	(1.09)	(1.17)	(1.07)
Sources	Treatment	Concentration		TXC	
SEM	0.002	0.001	0.004		
CD	0.010	0.010	0.020		

* Ö×+1 transformed values

Table 20.5: Percent Disease Incidence on Fruits (Pooled Data)

Tr. No.	Treatment details	Dosage (%)	Mean percent disease incidence on fruits after 9th spray	Mean of 4 th , 6th and 9th spray	C:B ratio
T1	Copper sulphate (CS)	0.1	21.25 (27.41)	31.07 (33.69)	1:3.4
T2	Soap nut (SN)	5	28.75 (32.40)	36.79 (37.26)	1:3.2
T3	5% Cu	0.3	21.25 (27.41)	29.29 (32.58)	1:3.3
T4	Pseudomonas fluorescens (P. f.)	1	18.75 (25.56)	27.68 (31.50)	1:4.8
T5	P. f. + SN	1+5	15.00 (22.51)	25.18 (29.65)	1:4.6
T6	P. f. + PPFM (strain 10)	1+1	26.25 (30.73)	34.11 (35.61)	1:2.8
T7	P. f. → SN → 5% Cu → P. f.→ COC → COC + SC → CS P. f → 5% Cu → SN	As per above treatments	36.25 (36.99)	43.39 (41.19)	1:3.7
T8	→ SN → 5% Cu → P. f. → SN → CS → COC + SC → 5% → Cu SN → P. f	As per above treatments	33.75 (35.49)	41.79 (40.23)	1:4.2
T9	COC+SC	0.3+0.05	11.25 (19.46)	21.43 (26.93)	1:5.3
T10	Control		52.75 (46.60)	51.25 (45.74)	
	SEm±		1.2		
	CD at 5 %		3.48		

21

Tracing the Evolutionary Origin of *Xanthomonas axonopodis* pv. *punicae* Causing Bacterial Blight of Pomegranate using Multilocus Sequence Typing

Aundy Kumar[1], Kalyan K. Mondal[1], and Jyotsana Sharma[2]

[1]Division of Plant Pathology, Indian Agricultural Research Institute,
New Delhi-110 012, India
[2]ICAR-National Research Centre on Pomegranate, Solapur-413 255, Maharashtra
E-mail: kumar@iari.res.in

Pomegranate (*Punica granatum* L.) is an ancient and important fruit crop of subtropical and tropical regions of the world. It is widely grown in Iran, India and Mediterranean regions of Asia, Africa and Europe. India is the largest pomegranate growing (0.193 million ha) and producing (2.2 million tonnes) country of the world. At the global level, India and Iran are the major producers of pomegranate collectively contributing about 85 percent of the total pomegranate production. Other important producers include Spain, Afghanistan, Pakistan, Egypt, Jordan, Tunisia, Lebanon, Israel, Chile, Peru and the US. Bacterial blight is one of the major production constraints in all pomegranate growing states of India such as Maharashtra, Karnataka, Andhra Pradesh, and Gujarat and to a small extent Rajasthan and Tamil Nadu. The disease has been recently reported from other parts of the world such as South Africa and Pakistan.

While India achieved a peak production of 8.84 lakh tonnes in 2007-08, the output has declined drastically due to bacterial blight disease which has negatively impacted the export where over 40percent reduction was experienced. Later, production rose to 17.89 lakh tonnes in 2014-15.

The causal bacterium was confirmed as *Xanthomonas axonopodis* based on biolog based bacterial identification system. The bacterium was characterized by a panel of phenotypic and genotypic methods in order to deduce its species identity and the consequent taxonomic position. The biolog based phenotypic fingerprinting reveals that the strain is close to *Xanthomonas axonopodis* pv. *juglandis* based on database search. Colonies on nutrient agar were yellow, raised, and translucent with smooth margins. *Xanthomonas axonopodis* pv. *punicae* associate phenotypes such as slow growth rate on nutritional media, sensitivity to salt above one percent, absence of growth above 35°C and accumulation of diffusible pigment called fuscan in the medium were documented. These phenotypes were exploited for identification of the bacterium. Identity of the bacterium as *Xanthomonas axonopodis* was further confirmed based on sequence comparison of 1500bp 16S rDNA by closest match with GenBank entries. However, the identity confirmation has not been done at strain level which is a prerequisite for establishing the etiology unambiguously. Being a ubiquitous bacterium in plant associated environment the strain level identity has become an essentiality. *Xanthomonas* spp. are known to cause different diseases on a broad range of diverse host plants including economically important ones such as rice, citrus, cabbage, and anthurium. Bacterial diseases are re-emerging worldwide owing to multitude of factors including changing agricultural practices instigated by anthropogenic reasons and the climate mediated reasons. Several new collections of bacterium causing leaf blight disease have been made from all pomegranate growing regions of Indian subcontinent.

The genomic era has witnessed a quantum jump in the full genome data being uploaded. Inspite of voluminous sequence information documented in public databases, molecular evolution, population genetics, ecology, and epidemiology of several plant-pathogenic and plantassociated bacteria are still poorly understood. Genome information and the associated genotypic tools, presently available, would help to unravel the evolution of the bacterial pathogens. Deciphering path of the bacterial strain evolution *i.e.*, diversifying, directional, or purifying, is a fundamental prerequisite for understanding the epidemiology and the pathogen migration across time, space (location) and crops (host). Genome wide analysis of bacterial pathogens by genotypic tools would be useful to this effect.

Multilocus Sequence Typing, a powerful genotypic technique for inferring evolutionary relationships at the inter-specific and intra-specific levels was adopted for the bacterium isolated from bacterial blight infected samples collected from several states such as Maharashtra, Andhra Pradesh, Karnataka, Himachal Pradesh and Delhi. The analysis specifically target the "core" genome when slowly evolving housekeeping genes encoding for proteins essential for the survival of microorganism are analyzed. Genome wide analysis of the plant pathogenic bacterium is essential to decipher its evolutionary dynamics. Housekeeping genes like *dnaK, fyuA, gyrB, rpoD, fusA, gltA, gapA, lepA* etc were amplified, sequenced and analysed. The gene sequences were analysed for Indels and SNPs and allelic variants were identified and assigned allele numbers. The analysis clearly revealed its genetic similarity with *Xanthomonas axonopodis*pv. *citri* (~*Xanthomonas citri* pv. citri) and *Xanthomonas axonopodis* pv. *malvacearum* (*Xanthomonas citri* pv. *malvacearum*) respectively, causing bacterial canker in citrus and bacterial blight in cotton. It may be speculated that the close proximity of these crops in the same ecological niche in central Indian subcontinent would have played a major role in selection of respective pathovars on these hosts. The relatively less genetic distance as observed in the MLST data is a pointer for such a probability.

We have analysed a "limited population" of the pathogen representing major pomegranate growing regions in India including the isolate collected in 1950s and preserved in National Collection of Plant Pathogenic Bacteria under FERA, UK. The data revealed that the isolates obtained from diverse geographical location and distant time points were identical as they shared same allele profiles. Identical allele profile found for the isolates further indicate its narrow genetic base or low genetic diversity and the strain responsible for bacterial blight epidemics in central India could be due to wide spread horizontal transmission of the bacterial pathogen from one place to another and passed on to seasons through, a safe carrier most likely, the vegetatively propagated planting material. Coupled with the practice of mono culturing of one or few elite cultivars for good harvest in vast stretch ofland, the pathogen has aggravated and accelerated the epidemics that we witnessed in our country. However, the study needs to be expanded to large collection of isolates *of Xanthomonas axonopodis pv. punicae* representing unconventional growing regions where the new disease outbreak has been observed.

22

Combating Pomegranate Bacterial Blight: Insights from Research Advances on Related Pathogen

Ginny Antony

Department of Plant Science, Central University of Kerala, Riverside Transit Campus, Nileswar, Kasaragod-671 314, Kerala

Bacterial Blight of Pomegranate is a major constraint on pomegranate production in India since 2005. In recent years bacterial blight of pomegranate has assumed epidemic proportions in northern Karnataka and Maharashtra seriously threatening its cultivation and has led to drastic fall in its productivity in the major pomegranate growing regions. This disease has also been reported from Pakistan and South Africa indicating that it has become an emerging threat to places where pomegranate is cropped. The draft genome sequence of the pathogen *Xanthomonas axonopodis* pv. *punicae (Xap)* was published in 2012 and revealed that *Xap* is closely related to *Xanthomonas axonopodis* pv. *citri,* the causal agent of citrus canker. Recent research conducted on related xanthomonads like *Xanthomonas oryzae* pv. *oryzae* and *Xanthomonas axonopodis* pv. *citri* with their host plants have revealed that single major effectors that belong to Transcription Activator Like effector (TALe) family targets specific host genes called Susceptiblity genes. The induction of S genes in these systems is critical for the virulence of the pathogen and hence identification of S genes presents an ex ellent opportunity for controlling these pathogens. Insights from these studies can be extrapolated to the pomegranate bacterial blight pathogen *Xap* to understand its pathogenicity mechanism.

Pomegranate (*Punica granatum* L.) is a high value crop. Last decade has seen a surge in the demand of this fruit in the international market and therefore had brought in huge dividends for pomegranate growers. Due to the high remunerative nature of the crop, there has been a steady increase in the area and production of this fruit crop in the country. Pomegranate being well suited to arid and semi arid regions has the potential to develop wastelands into farmlands. In India pomegranate is largely grown in Maharashtra, Karnataka, Gujarat, Andhra Pradesh, Rajasthan and Tamil Nadu. In the last 2-3 years pomegranate production in Maharashtra and Karnataka have come down drastically and its export to international markets have consistently dropped due to the high incidence of a deadly disease called Bacterial Blight. Bacterial Blight is caused by *Xanthomonas axonopodis* pv. *punicae* (*Xap*).

This disease assumed economic importance only recently and therefore very little research efforts have been put into it. All commercially grown pomegranate cultivars are susceptible to this disease with no incidence of resistance reported so far. The draft genome sequence of the pathogen *Xanthomonas axonopodis* pv. *punicae* (*Xap)* strain LMG 859 was published in 2012 (Sharma, V. *et al*, 2012) and revealed that *Xap* is closely related to *Xanthomonas axonopodis* pv. *citri* (Xac*)*, the causal agent of citrus canker. The genus *Xanthomonas* belong to gammaproteobacteria (Parkinson et al, 2007) and has 27 species which are found to infect ~400 different plants including commercially important crops like rice, banana, citrus, cassava, tomato and bean. Each species has multiple pathovars and show high degree of host specificity. They gain entry into the plant through wounds or natural openings like hydathodes or stomata and colonize the xylem vessels or the intercellular spaces of the mesophyll parenchyma tissue. To date complete genome sequences of eleven different strains of *Xanthomonas* and draft sequences of eight other strains including LMG 859 strain of *Xanthomonas axonopodis* pv. *punicae* is available (Ryan P.R. *et a.l*, 2011; Sharma, V. *et al.*, 2012). This enormous wealth of sequence data can reveal the secrets behind the extraordinary diversity and pathogenic adaptations of this genus. The sequenced genomes of *Xanthomonas* spp. range in size from 3.77Mb to 5.42 Mb and are predicted to encode ~ 4000 proteins. Some of these pathogens like *Xanthomonas citri* and *Xanthomonas euvesicatoria* carry multiple plasmids that encode genes for virulence. Virulence factors of *Xanthomonas* spp. include toxins, adhesions, lipopolysaccharides, gum genes, type II secretion system and type III secretion system and its associated effectors. All characterized *Xanthomonas* spp. encode type II secretion system and most *Xanthomonas* spp. possesses type III secretion system. Type III secretion system along with its effectors are critical players in the virulence of most *Xanthomonas* spp. A total of 52 type III effector families are identified and a few accessory proteins called harpins that assist in the translocation of effector proteins. Type III effectors are categorized into TAL effectors and non TAL effectors. TAL effectors from *Xanthomonas oryzae* pv.*oryzae* (Xoo) is well studied and are found to play major role in the virulence of the strains that carry them and also determine host specificity. TAL effectors are highly conserved nuclear localized proteins with a central repeat region that directly interact with host gene promoter and transcriptionally activate them. Each sequenced Xoo strain harbor multiple TAL effectors (15-19) and are considered as hot spots of recombination that enables adaptation and evolution of new pathogen races. Knock out and complementation studies on TALes suggested that each strain depend on a single or in some cases two TAL effectors to promote disease in the host. These TAL effectors are referred to as major TALes. Knock out mutants of major TALes are avirulent or weakly virulent (Yang and White, 2004; Yang, B. *et al.*, 2006; Antony, G.,*et al.*, 2010). S gene induction by Xoo is critical for successful rice colonization and host susceptibility. Silencing of two of these S genes using RNAi resulted in plants that are completely resistant to the respective Xoo strains. So far 3 different susceptibility genes and their cognate effectors have been identified (Yang, B.*et al.*, 2006; Antony, G., *et al.*, 2010; Liu, *et al*, 2011). All the three S genes identified belong to the MTN3 gene family and rice genome has ~17 MTN3 genes scattered throughout the genome. Major TAL effectors interact with their cognate S genes in a gene for gene manner and bind to specific elements in their promoter called Effector Binding Elements (EBE) (Antony, G. *et al*, 2010). TALes follow a simple code for DNA recognition and binding. TALes are highly conserved proteins with a series of nearly perfect repeats at their center, the exception being two amino acids at the 12th and 13th position of each repeat called repeat variable diresidues (RVDs). Each RVD specifies a single bp on the contiguous EBE and the length of EBE is determined by the number of central repeats (Boch, J., *et al.*, 2009; Moscou and Bogdanove, 2009). Recent studies indicate that S genes *Os8N3* and *Os11N3* function in the efflux of glucose from cells and pathogen hijacks the transcription of these genes to enhance nutrient supply in the intercellular spaces where they colonize (Chen L.Q. *et al.*, 2010). Owing to their cellular role as sugar transporters these S genes are also called SWEET genes. Related effector PthA from *Xanthomonas axonopodis* pv. *citri* (Xac) was the first type III secreted TAL effector demonstrated as essential for virulence as well as the development of pustule formation, typical symptom of citrus bacterial canker (Swarup, S. *et al.*, 1992). Subsequently genes for the TAL effectors PthA4, PthAw, PthB, and PthC from genetically diverse Xanthomonas strains that cause bacterial canker were found to be involved in pustule formation (Al –Saadi A. *et al.*, 2007). A recent study has reported that these TAL genes promote pustule formation by the induction of a transcription factor Lateral organs boundaries 1 in citrus and hence *CsLOB-1* is the disease susceptibility gene for citrus canker (Hu, Y. *et al.*, 2013). Exact function of *CsLOB-1* in the host cell is yet to be determined. Non TAL effectors also called *Xanthomonas* outer proteins or Xop effectors also play major role in the virulence. Xap *Xop N* was recently demonstrated to modulate bacterial growth in planta and suppress basal defense responses of the plant like callose deposition during infection (Kumar R and Mondal K.K, 2013).

The aforementioned studies validate the importance of type III effectors in the virulence of xanthomonads. In rice and citrus, effector proteins confer specificity at the levels of pathogen race and host cultivar. For this reason, the characterization of the effector proteins, the identification of their molecular targets in the plant is absolutely essential. The effectors can then be manipulated to modulate host susceptibility or resistance. Understanding the host susceptibility mechanism is vital in the case of pomegranate bacterial blight as no source of resistance to this deadly disease have been reported or identified.Comparative genomic approaches as well as transcriptional profiling of host responses in pomegranate may enable the identification of key determinants of host specificity and susceptibility.

23

Recent Insight into the Molecular Basis of Pathogenesis by *Xanthomonas axonopodis* pv. *punicae* in Pomegranate

Kalyan K. Mondal

Division of Plant Pathology,
Indian Agricultural Research Institute, New Delhi-110 012
E-mail: kkmondal@iari.res.in

Introduction

Pomegranate (*Punica granatum* L.) is an important fruit crop of India with export pot ential. Bacterial blight caused by *Xanthomonas axonopodis* pv. *punicae (Xap)* is a major biotic threat to pomegranate cultivation in India (Mondal and Mani, 2009; Mondal *et al.*, 2012a). Xap is closely related to *X. citri* subsp. *citri* as well as subsp. *malvacearum* (Mondal *et al.*, 2012b). Xap genome of 4.94-Mb contains 4,385 predicted coding regions (CDSs) and 50 tRNA and 3 rRNA genes. The G+C content in the Xap genome is 64.9 percent (Sharma *et al.*, 2012). The disease management through application of bactericides, streptomycin and copper based fungicides have limited success owing to their cost and inconsistent performance against the existing strains. New strategies including biological control and new generation molecules like nano copper has shown some promise (Mondal and Mani, 2012). However, novel strategies, targeting suppression of pathogen's virulence genes as well as host susceptibility genes, requires detailed understanding on pomegranate-bacteria interaction. Thus, the recent efforts are geared up on biochemical and molecular studies targeting pathogens virulence factors like type III effectors (T3Es). In this direction, the present paper discusses few leads with respect to the precise role of T3Es of Xap in governing virulence and inducing blight in pomegranate.

Materials and Methods

Bacterial Strain: All the experiments were carried out with a rifampicin resistant strain ITCCBD 0003 of Xap, referred as Xap-Rif[100].

Phylogenetic Analysis: For phylogenetic analysis, the full length nucleotide sequence of *xopN* from *Xap* was aligned with 10 *xopN* sequences from different *Xanthomonas* strains using alignment tool Unipro U gene (1.11 version). The phylogenetic tree and similarity matrix was based on pairwise alignment constructed with the help of Mega 5 software.

Construction of Mutant for *xopN*: XopN deletion mutant was developed employingdouble crossing over based recombination strategy (Kumar and Mondal, 2013).

***In planta* Assay:** The pomegranate plant (cv. Bhagwa) was used for virulence as well as growth assay. Cell suspension (10^7 cfu ml^{-1}) of wild, mutant Xap$\Delta xopN$, or the complemented Xap$\Delta xopN$ + *xopN* was infiltrated. The appearance of intense water-soaked lesions in the infiltrated areas seven days after infiltration was indicative of the virulent phenotype. For bacterial growth, bacterial cell count was estimated at 0h, 24 h, 48 h, 72 h, and 96 h intervals of post infiltration. For callose assay, the Xap-infiltrated pomegranate leaves were harvested 12 h after bacterial infiltration and were immersed in 5 ml of alcoholic lactophenol solution before incubation for 30 min at 65ºC to remove chlorophyll. The cleared leaves were stained with 0.01percent aniline blue after rinsing thoroughly in 50percent ethanol and in water.

Results and Discussion

Like other phytopathogenic *Xanthomonas*, Xapsecretes type III effectors (T3Es) directly into the pomegranate cells through the type III secretion system (T3SS). The T3Es play important role in governing bacterial virulence and suppress pathogen-associated molecular pattern (PAMP)-triggered plant immunity (PTI) (Mondal *et al.*, 2014). Thus T3Es modulates the blight symptoms in pomegranate. *Xap* contains six T3Esof *Xanthomonas* outer proteins (Xop) family, namely XopC2, XopE1, XopL, XopN, XopQ and XopZ (Kumar and Mondal, 2013).

The detailed investigation on XopN, a conserved effector in *Xanthomonas*, indicated that XopN contributes significantly to the bacterial growth, and virulence. The XopN shared 98.6percent sequence identity with pathovar *citri*. Our recent study indicated that the all the T3Es of Xapare regulated via *hrp*-dependent manner and their expression was induced under XOM medium.

The pomegranate leaves upon infiltration with Xap*ΔxopN* did not produce prominent water-soaking symptom indicating the less-virulent phenotype of the mutant. The absence of intense water-soaking on leaf infiltrated with the Xap*ΔxopN* further confirmed the role of the XopNeffectorin contributing to virulence. The suppression of virulence due to loss of T3SS effectors has been documented in many phytopathogenic bacteria including *Xanthomonas campestris* pv. *vesicatoria* infecting tomato and *Xanthomonas oryzae* pv. *oryzae* infecting rice (Jiang*et al.*, 2008; Kim *et al.*, 2009; Metz *et al.*, 2005; Song *et al.*, 2010). Likewise, our study first time demonstrated the role of *xopN* T3SS effector of Xap infecting pomegranate in the suppression of cell-wall-associated immune response.

Plant responds to bacterial infection by depositing callose, a β-1, 3-glucan, around the site of inoculation. Thus, callose deposition is considered as a basal defense response associated with PTI. We observed that pomegranate leaves inoculated with Xap *ΔxopN* had morecallose depositions (209.6 callose deposits per leaf disk) than leaves inoculated with wild type Xap (73.8 callose deposits per leaf disk). The data indicates that XopN is required to suppress Xap-induced callose deposition in pomegranate leaves. Many T3Es including XopN that are known to suppress the PTI and their role in PTI suppression has been demonstrated through less callose deposition, an early event in plant cell-wall-associated defense reaction (Grant *et al.*, 2006; Mudgett, 2005).

The cell-death assay on *Nicotiana glutinosa* using combination of PTI-inducer and challenger strains indicated that the susceptible reaction (SR) was compromised in tobacco leaf co-infiltrated with *Xap ΔxopN* (challenger) and *Pseudomonas aeruginosa* (inducer) resulting in hypersensitive response (HR).

Our investigation on XopN effector clearly suggests that Xap employs the T3Es to induce blight in pomegranate and XopN contributes substantially in governing virulence to the Xap. However, the detailed functions of the other T3Es in bacterial virulence need to be investigated. This insight would lead to a better understanding on T3Es-mediated pathogenesis and to explore sensible strategies towards managing the blight bacterium.

24

Pomegranate Diseases and their Management

Jyotsana Sharma and K.K. Sharma

ICAR-National Research Centre on Pomegranate, Solapur- 413 255, Maharashtra
Email: jyotisharma128@yahoo.com

Introduction

Pomegranate (*Punica granatum* L.) is an important fruit crop of Maharashtra. Being a highly remunerative crop, it is gaining popularity in other states of India also. *A decade ago no major diseases causing economic losses were reported on pomegranate.* With remunerative returns from small area and the growing demand in export and local market, growers adapted new improved varieties and hi-tech horticulture. As a result some of the diseases which were practically unknown or of little economic importance are reported to be a serious problem, today. Among various diseases of pomegranate, bacterial blight/oily spot (*Xanthomonas axonopodis* pv. *punicae)* and wilt/decline (*Ceratocystis fimbriata* major cause along with root knot nematode -*Meloidogyne incognita*) were, till recently the two most important diseases that resulted in economic losses to the growers, in all major pomegranate growing areas of India. With increasing interest of the growers in the Dollar crop and climate changes, several fungal fruit spots, rots and scab which cause qualitative as well as quantitative losses are also becoming cause of concern. Different researchers have attributed various pathological causes to these diseases. Predominant fruit spots are caused by *Cercopsora punicae* and *Alternaria alternata,* scab (*Sphaceloma punicae), a*nthracnose (*Colletotrichum gloeosporioides)* and fruit rots by *Colletotrichum gloeosporioides* and

Phytophthora nicotiane. These pathogens also result in foliage infections, but are manageable. Among postharvest rots from different states in India, about 20 species including 10 genera have been reviewed. The major rots include species of *Aspergillus, Cephalosporium, Cladosporium, Curvularia, Glomerella, Paceilomyces, Penicillium, Rhizopus, Sclerotium and, Phomopsis,* however, losses due to postharvest rots are very low. Only few reports are available on virus or virus like organism (VLO) associated with pomegranate and that too, not from India, however, economic losses have not been reported due to these diseases. In all major pomegranate diseases prophylactic measures are better than cure, as therapeutic treatments after onset of diseases are less promising. Integrated disease and insect pest management is the best and most economical option for economic gains. The various components of IDM used for managing important diseases are discussed here.

Management

Prophylactic measures for managing most plant problems are most economical and effective. Integrated Disease and Insect Pest Management (IDIPM) in pomegranate is a step in this direction.

The various components of IDM used for managing important diseases are discussed here.

Site Selection

Wilt-infested sites should be avoided or properly sterilized using recommended chemicals like formalin @ 2.5-5 percent before planting new orchard. Avoid deep heavy soils with poor drainage, select site having light-medium soil for establishing new orchard.

Sanitation

a. Clean Planting Material: New planting should be done with certified disease free planting material or tissue culture saplings. The pomegranate being propagated by vegetative means (air layers and hard wood cuttings) planting material plays a significant role in the introduction of diseases like bacterial blight and wilt in new areas. The wilt organisms especially fungi and nematodes are easily transmitted in cuttings/air layers and through soil used for establishing the saplings. Unrestricted movement of cuttings or other propagative material is potentially dangerous in introducing the disease in new areas. Hence, do not prepare air layers or cuttings from apparently healthy shoots arising from infected plant. The soil used for bagging the planting material should be sterilized using chemicals or through solarization.

b. Barrier Crop: Plant few rows/a of wind breaks like *Casuarina* (Saru), *Greviellea robusta* (Silver oak), *Sesbania grandiflora, Musa indica (banana) or any other tall trees with dense foliage* around the orchard to act as barrier crops for air borne diseases like bacterial blight and fungal spots. It has been observed that pomegranate plantations surrounded by barrier crops with dense plantation generally have no/low disease in spite of infected orchard in the neighbourhood.

c. Orchard Sanitation: Orchard sanitation plays an important role in not only reduction of bacterial blight but also wilt and other diseases. Infected plant parts (leaves, flowers, fruits & twigs) should not be left in the orchards nor dumped near orchard, nor thrown in irrigation channels. The orchard should be swept clean to collect all fallen plant parts and burnt. Ground below the canopy in the basin of tree should be drenched @ 25 kg/1000 lwater/ha thrice a year. This will kill the pathogen inoculum on left over plant debris if any in the orchard. Orchard should be kept free from weeds, which may be latent carriers or multiplication ground for bacterial blight, nematodes and several other pathogens and insect pests. The affected dead wilted plants should be uprooted and burnt immediately and not dumped in/near the orchard for use at a later stage for firewood. While removing the wilted plants from the orchard for burning, protect the entire root zone with cover, as the propagules of the pathogen are present abundantly on the roots and they may spread to other healthy plants.

Once wilt disease is detected in the orchard, create a buffer zone- *i.e.* dig a trench about 3-4 feet between the wilted and healthy tree to avoid transmission to healthy trees.

Cultural Practices

a. Planting: Planting of pomegranate should be done at a plant to plant spacing of 3m and row to row spacing of 4.5m to avoid natural root or foliar contacts. Pits should be dug at least a month prior to planting and kept open to disinfect the pits by intense solar radiation during the day-preferably during hottest summer months. Just before filling pits drench the bottom and sides of the pit with 4-5 litres of 0.4% chlorpyriphos 20EC solution. Dust the pits with bleaching powder (a.i.33% Cl) @ 100g/pit before filling. Fill the pits with soil having, sand/murrum, silt and clay in equal proportions. In each pit mix the following in the top soil (30-50 cm): 10kg of FYM, 1 kg vermicompost, 0.5 kg Neem cake and 25g each of beneficial microflora like *Trichoderma* formulation, Phosphate Solublising Bacteria (PSB), *Pseudomonas fluorescens, Azotobacter* formulation, *Azospirillum* formulation, *Aspergillus niger* formulation and AMF. Alternatively the rows/ridges for planting should be prepared in advance, and solarized by covering with 25-75 micron transparent polyethylene mulch for 6 weeks in hottest summer months. This gives added advantages of increasing beneficial microflora population and killing soil borne pathogens, weed seeds and insect pest eggs. Plant the orchard with disease free plants.

b. Change of Season: In bacterial blight prone areas only *hasta bahar* or late *hasta bahar* crop must be regulated. This also reduces infection from other fungal spots and rots. In bacterial blight free areas, growers may take *bahar* convenient to them. Bacterial blight and most of the fungal spots, scab anthracnose and rots spread more, progress at an increased infection rate and are very severe in rainy season crop, hence, in areas where blight or other diseases are a problem, flower regulation time should be shifted from June-July to September- October/November to take rabi/*hasta*/late *hasta bahar* crop specially in Maharashtra, Karnataka and Andhra Pradesh. Pomegranate is hardy crop and can withstand diverse weather conditions, hence in areas with severe winters though the winter crop can be taken if disease problems are there, but proper irrigation will be required to prevent fruit cracking and crop may take longer time for maturity.

c. Nutrient Management: The balanced supply of mineral nutrients increases inherent tolerance to diseases and escape from diseases, hastening or delaying maturity, enhancing physiological resistance. They reduce pathogenic activity and its virulence. Incorporation of organic manures like vermicompost and neem cake during the rest period or prior to flowering and application of macro- and micronutrients during different fruit development stages improve plant health, quality produce and result in reduction of bacterial blight and other diseases. Use of salicylic acid @0.3g/l is provides resistance to pomegranate against blight. Nitrogen excesses or deficiencies, both lead to increased severity of most foliage diseases. Potassium, calcium, magnesium, manganese, zinc, boron, copper are known to play a positive role in disease management.

d. Irrigation: Crop should be irrigated depending on plant age, growth stage and season. Almost all pomegranate diseases increase with high humidity in soil and the microclimate around the tree. During fruit bearing stage regular irrigation should be provided. During rest irrigation should be given only for its survival and proper health. Avoid water stress or overwatering that may result in fruit cracking and aggravating diseases like wilt. The drippers should be kept 6-12" away from the stems depending on plant size.

e. Intercrops: Crops like onion, tomato, chili, potato, capsicum, gram, legumes, cucurbits, gerbera, gladiolus *etc.* aggravate nematode infestation and hence should be avoided as intercrop. Cucurbit crops increase insect problems also, which in turn transmit and spread several pathogens, hence should be avoided. Sunhemp and other green manure crops help improve natural beneficial microflora and should be preferred as intercrop. Maize, wheat, sorghum, bajra, mustard also reduce nematode population. Planting *Tagetes erecta* (African marigold) varieties – 'Pusa Basanti Gainda' and 'Pusa Narangi Gainda' for continuously 6-7 months is beneficial for reducing nematode population in infested orchards. For effective results these should be grown for at least 6-7 months continuously.

F. Pruning: Proper pruning and training should be followed, to develop optimum canopy and to avoid contact of branches with neighbouring plants. In orchards where severe bacterial blight infection is noticed, heavy pruning immediately after harvest must be done to remove all stems with fresh blight infections/cankers, as bacterial blight is not a systemic disease. Infected stems must be pruned about 2" below the infected area. Cankers, if not removed due to any reason, should be pasted/painted with a bactericidal paint like Bordeaux paste (10%), copper oxychloride paint, Chaubattia paste. Pruning tools – secateurs etc should be sterilized after handling each infected tree with sodium hypochlorite (2.5%). Any severely infected plant due to blight must be uprooted burnt and replaced with new disease free plant or cut from base 2-3 inches above ground level. New well growing sprouts should be trained for new disease free plant. Bordeaux mixture (1%) should be sprayed immediately after pruning. Wilt affected plants should be treated as soon as first signs are observed. If the plant does not recover, it is advisable to uproot and burn the plant, so as to avoid contamination of healthy plants through pruning. Trees with more than 30% canopy loss due to wilt should not be treated; they should be uprooted and burnt.

Host Resistance

Development of resistant variety is the only long term solution for managing the diseases like BB and wilt. Work on breeding bacterial blight resistant/tolerant varieties is being given top priority in ICAR institutes. A collection of 236 accessions, including 177 indigenous and 59 exotic in the field gene bank at NRCP Solapur have been screened against bacterial blight, none showed resistance to bacterial blight. Unfortunately none of the pomegranate germplasm available in India, including the commercial cultivars has any resistance to blight, although some degree of resistance to blight has been found in a wild variety 'Daru' and a dwarf variety 'Nana'. The hybrids developed using these tolerant varieties at IIHR, Bengaluru and screened at NRCP, Solapur, have not yielded any resistant variety so far. All commercial pomegranate varieties grown in India, viz., 'Ganesh', 'Bhagawa', 'Arakta' and 'Mridula' are also susceptible to wilt pathogens. More than 50 germplasm screened for *C. fimbriata* wilt have not yielded any resistance.

Biological Control

Biological formulations if used should be reliable, fresh and used during rest period when no other fungicides/bactericides are used. Many bioagents give effective control of bacterial blight in winter season crop, however, none was found effective in rainy season. *Bacillus subtilis, Pseudomonas fluorescens* algal formulations, Neem Seek Extract @ 7.5 percent gave significant control of bacterial blight in field trials.

The soil application of *Bacillus subtilis, Paecilomyces lilacinus, Pseudomonas fluorescens, Trichoderma harzianum, Aspergillus niger* 10-15g/plant along with well-decomposed farm yard manure around the trunk of pomegranate trees helps to prevent wilt infections. Neem cake @ 2-3 kg/plant effectively checks incidence of wilt complex. At NRCP biofertilizer – Kalisena SA having Aspergillus niger - used against several soil borne pathogens causing wilt diseases of vegetable and plantation crops- was evaluated against *C. fimbriata,* causing wilt of pomegranate. Kalisena treatment kept wilt under check for 62 weeks, in comparison to control when wilt started after 12 weeks only, however maximum wilt control (75%) was in the combination treatments of Kalisena and a VAM preparation with *Glomus* sp.

Organic soil treatments with neem (*Azadirachta indica*), Kranj (*Pongamia pinnata*), Mahua (*Bassia latifolia*) and castor (*Ricinus communis)* cakes at 2.5 t/ha are effective in the management of root knot nematode.

Chemical Control

Bacterial diseases among all plant diseases are difficult to manage by chemicals alone. Bordeaux mixture, copper compounds, 2-bromo, 2-nitro propane-1, 3-diol (Bronopol) and the antibiotic streptocycline are the only chemicals which satisfactorily manage the disease under low disease pressure and environmental conditions which do not favour rapid disease development. Spray schedule comprising of bordeaux mixture (0.5-1%), streptocycline (500ppm), 2-bromo-2-nitropropane-1,3-diol@ 500 ppm alone or in combination with fungicides like copper oxychloride (0.25%)/ copper hydroxide (0.2%), carbendazim (0.1%) at 10-15-day interval reduce blight and increase yield of quality fruits. As far as possible chemical sprays should be combined depending on compatibility, as sprays without bactericide increase bacterial blight.

On observing first symptoms of wilt due to fungal pathogens in the orchard immediately drench soil with chlorpyriphos 20EC (0.25-4%) + carbendazim 50WP (0.2%) or propiconazole 25EC (0.2%) use 5-8 l solution/tree. Also drench at least 2-3 healthy plants on all the four sides around the infected plant/s, repeat the drenching three times at 20-25 days interval. As preventive measure, if bioagents have not been added, entire orchard can be drenched three times in a year at four months interval through drip. Soil sterilization prior to planting with formalin (2.5-5%) also controls wilt disease. Drenching with metalaxyl or dithane M-45 will be beneficial if *Phytophthora* is causing any loss.

For controlling shot hole borer (*Xyleborus* spp.) or stem borer which is associated with wilt disease, 10 litres preparation containing red soil (4kg) + Chlorpyriphos 20EC (20ml) + Copper oxychloride (25 g) needs to be applied on plant base up to 1-2 ft. from second year onwards twice a year. To control stem borer, inject in the holes on the trunk with DDVP 2-3 ml and plug the holes with mud. Where nematodes are a problem apply phorate 10G @10-20g/plant or carbofuran 3G @ 20-40g/plant, in the plant basin, in a ring near root zone and cover it with soil, three times a year at 4 month interval.

Leaf and fruit spots as well as rots caused by different fungal pathogens are effectively managed by the sprays of carbendazim (0.2%), thiophanate methyl (0.15%), mancozeb (0.2%), captan (0.2%), benomyl (0.2%) and copper oxychloride (0.25%) under field conditions. Sprays should start at pre flowering/flowering and continue after 10-15 days depending on weather conditions and nature of fungicide. Alternating sprays of chlorothalonil 75%WP (2g/l), formulation containing - tricyclazole 18% + mancozeb 62% WP @ 2-2.5g/l and propiconazole (1ml/l) after appearance of symptoms gives effective check of fungal rots.

For *Phytophthora* blight, spray the crop with formulation containing metalaxyl 8% + mancozeb 64% @ 0.25% or mancozeb (0.25%) or dimethomorph (0.1%) as soon as disease appears or as prophylactic in orchards where it makes yearly appearance. Fruit dip with dithane Z-78 (0.1%), captan (0.2%), benomyl (0.2%) are known to reduce post harvest rots.

The sap sucking insects – thrips, aphids, mites etc, leaf eating caterpillars and other insect pests should be kept under check by use of chemical/organic insecticides, as these may spread disease organism from one to other plant.

- The major steps of IDIPM for managing pomegranate diseases are -Planting and establishing new orchards at proper spacing with disease free planting material from a certified nursery.

- Providing balanced nutrition to plants, as nutrients play significant role in disease development and resistance.

- Avoiding *mrig bahar* crop at least for few years in disease prone areas with heavy summer rains, to reduce the inoculum build up.

- Adopting stringent orchard sanitation measures.

- Pruning blight infected stems as far as possible, painting them with recommended bactericidal paints, uprooting severely cankered and wilted/ dead plants and burning them.

- Apart from recommended sprays during crop season, taking prophylactic sprays of bordeaux mixture (1%) altered with bactericide bronopol + broad spectrum fungicide at 15 days interval depending on disease present in the orchard or neighbouring orchards, after harvest, during rest period and also after planting a new orchard. Wilt treatments should be taken up on observing the first signs.

- Bioformulations may be applied to soil during rest period if chemicals are not being used.

- During crop season take only need based sprays at recommended doses, too many sprays increase the bacterial blight disease if the spray chemical is ineffective against the pathogen by providing much needed water to the pathogen.

- Using good quality non-ionic spreader sticker with sprays except with bordeaux mixture. It improves efficacy of the spray chemical as pomegranate foliage and fruit surfaces are glossy.

- Following rest period of 3-4 months and taking only 1 crop in a year to improve plant vigour.

- The benefits from following integrated management schedule have been well demonstrated in pomegranate growing areas of Maharashtra, Karnataka and Andhra Pradesh through multilocation trials and well documented.

25

An Assessment of Economic Losses due to Oily Spot Disease of Pomegranate in Western Maharashtra

V.A. Shinde[1], V.M. Amrutsagar[2] and G.K. Bembalkar[3]

Zonal Agricultural Research Station, Solapur-413 002, Maharashtra
(Mahatma Phule Krishi Vidyapeeth, Rahuri)

Introduction

Pomegranate is the most important fruit crop of the tropical and sub-tropical region. Pomegranate is cultivated throughout the country of which the leading states are Maharashtra, Karnataka, Andhra Pradesh, Gujarat, and Rajasthan. The area under pomegranate in India is 1.93 lakh ha (2015-16) out of which nearly 60 per cent (98000 ha) area is in Maharashtra. In Maharashtra, Solapur and Nasik are the leading districts in pomegranate cultivation comprising 36.73 and 28.93 per cent area under pomegranate, respectively. Next to these two districts, the pomegranate is grown in Ahmednagar (8.14%), Pune (5.46 %) and Sangli (6.17 %) districts. Maharashtra contributes about 75 per cent (11.92 lakh tons) of the total production of India (17.98 lakh tons). However, the oily spot disease was reported in Solapur, Nasik and also some parts the state from the year 2002 and the heavy losses in the yield of pomegranate were noticed. The economic losses thus reported were in the range from 1 per cent to 100 per cent. In view of this, the present investigation was undertaken in order to know the extent of economic losses due to oily spot disease. Accordingly, the present study was undertaken with the following objectives.

Objectives

1. To ascertain the physical and economic losses in terms of yield and returns of pomegranate due to oily spot disease.

2. To study the awareness and adoption of control measures for oily spot disease.

Methodology

The districts viz; Solapur and Nasik districts having the maximum acreages under pomegranate were selected purposively. From the above districts, Sangola and Pandharpur tahsils from Solapur district and Deola from Nasik district were selected as the pomegranate orchards from these tahsils were affected by the oily spot disease on a large scale. From each tahsil, two villages in which large number of pomegranate orchards are affected by oily spot disease were selected for the study and thereby; an all four villages were selected. From each village, twenty pomegranate growers were selected. Thus the total sample comprised of one hundred twenty cultivators. The required information was collected by survey method with the help of specially designed schedules. The data were analyzed with the help of tabular method and by using suitable statistical tools.

Results

Area Under Pomegranate and the Proportion of Affected Orchards on Sample Farms

The information on varietywise area under pomegranate and affected orchards on sample farms is presented in table 25.1.

Table 25.1: Per Farm Varietywise Area Under Pomegranate and Affected Orchards

Sr. No.	Particulars	Solapur	Nasik	Overall
1.	Total holding	3.29	2.47	3.01
2.	Fallow land	0.05	0.09	0.06
3.	Cultivable area	3.24	2.38	2.95
	a) Unirrigated	0.69	0.28	0.55
	b) Irrigated	2.55	2.10	2.40
4.	Area under pomegranate	1.29 (100)	1.34 (100)	1.31 (100)
	a) Ganesh	0.68 (52.71)	0.12 (8.95)	0.49 (37.40)

Contd...

Sr. No.	Particulars	Solapur	Nasik	Overall
	b) Bhagwa	0.59 (45.74)	0.68 (50.75)	0.62 (47.33)
	c) Mrudula	0.01 (0.78)	0.11 (8.21)	0.04 (3.05)
	d) Arakta	0.01 (0.78)	0.43 (32.09)	0.15 (11.45)
5.	Affected area	0.13 (10.08)	0.16 (11.94)	0.14 (10.69)
	a) Ganesh	0.07 (5.42)	0.01 (0.75)	0.05 (3.82)
	b) Bhagwa	0.06 (4.65)	0.08 (5.97)	0.07 (5.34)
	c) Mrudula	-	0.01 (0.75)	-
	d) Arakta	-	0.06 (4.48)	0.02 (1.53)

Figures in the parentheses are the percentages to the area under pomegranate

The average per farm land owned by the sample pomegranate growers was 3.01 ha., at the overall level and it was 3.29 ha. and 2.47 ha. in Solapur and Nasik districts, respectively. Of this the area under pomegranate was 1.31 ha., at the overall level and it was 1.29 ha. and 1.34 ha. in Solapur and Nasik districts, respectively. The proportion of area under different varieties of pomegranate viz; Ganesh, Bhagwa, Mrudula and Arakta on the sample farms was 37.40, 47.33, 3.05 and 11.45 percent, respectively.

The proportion of area under different varieties of pomegranate viz; Ganesh, Bhagwa, Mrudula and Arakta in Solapur district was 52.71, 45.74, 0.78 and 0.78 per cent, respectively. While, in Nasik district the proportion of area under different varieties of pomegranate viz; Ganesh, Bhagwa, Mrudula and Arakta was 8.95, 50.75, 8.21 and 32.09 per cent, respectively.

The proportion of area under pomegranate affected by the oily spot disease was 10.69 per cent of the total area under pomegranate on the sample farms. While it was 10.08 and 11.94 per cent in Solapur and Nasik districts, respectively. The varietywise area affected by the oily spot disease was 3.82,5.34,0.00 and 1.53 per cent of the Ganesh, Bhagwa, Mrudula and Arakta varieties, respectively.

Variety Wise and District Wise Distribution of Total Area Affected by Oily Spot Disease

The information on variety wise and district wise distribution of total area affected by oily spot disease of the sample cultivators is given in table 25.2.

Table 25.2: Varietywise and Districtwise Distribution of Area Affected by Oily Spot Disease

Sr No.	Variety	District	Area	Affected area	Per cent
1.	Ganesh	Nasik	7.36	0.81	11.00
		Solapur	40.69	4.31	10.59
		Total	48.05	5.08	10.57
2.	Bhagwa	Nasik	36.65	4.62	12.61
		Solapur	33.3	3.66	10.99
		Total	69.95	8.28	11.84
3.	Arakta	Nasik	25.85	3.49	13.5
		Solapur	0.75	0	0
		Total	26.60	3.49	13.12
4.	Mrudula	Nasik	6.9	0.62	8.98
		Solapur	0.70	0.02	2.86
		Total	7.60	0.64	8.42
		Overall	152.20	17.49	11.49

It was observed from the table that, within the four varieties, maximum (13.12%) area was affected of Arakta variety followed by Bhagwa (11.84 %), Ganesh (10.57 %) and Mrudula (8.42 %) varieties.

The districtwise analysis revealed that the Ambe baher taken in case of Bhagwa variety in Nasik district, the percentage of affected area was nearly double than affected area under Solapur district, in remaining bahars no remarkable difference was observed in both the districts.

Bahar Wise Distribution of Area Affected by Oily Spot Disease

The *bahar* wise area affected due to oily spot disease is presented in the table 25.3.

Table 25.3: *Bahar* Wise Distribution of Affected Area (Ha.)

Sr. No.	Bahar	Total Area	Affected area	Percentage of affected area
1.	Mrug	78.52	8.36	10.65
2.	Hast	1.00	0.04	4.00
3.	Ambe	70.68	8.65	12.52
4.	Overall	152.20	17.49	11.49

From the table it was revealed that 11.49 per cent area was affected by the oily spot disease. However, within the three *bahars* maximum percentage was noticed in *Ambe bahar* (12.52 %) followed by *Mrig* (10.65 %) and *Hast bahars* (9.22 %).

Bahar Wise and Plant Density Wise Distribution of Area Affected by Oily Spot Disease

The bahar wise and plant density wise area affected due to oily spot disease is presented in the table 25.4.

Planting at proper distance is very much essential for better yield and to have good aeration in the garden, which ultimately reduces the chances of congenial atmosphere for microbes responsible for oily spot, and other diseases that's why the plant densitywise analysis was carried out. It was observed that, area affected due to oily spot disease was maximum in case of plant density above 926 especially in Mrug bahar. However, there was no any sample orchard for which recommended planting distance was adopted.

Table 25.4: *Bahar* Wise and Plant Densitywise Distribution of Area Affected on Sample Farms

Sr. No.	Plant density range	Bahar	No of affected plants	Area	Affected area	Percentage of affected area
1.	≤ 554	*Mrig*	14	15.50	1.33	8.58
		Ambe	4	8.60	0.85	9.88
2.	555 to 739	*Mrig*	7	10.72	1.28	11.94
		Hast	1	0.40	0.03	7.50
		Ambe	14	12.65	1.23	9.72
3.	740	-	-	-	-	-
4.	741 to 925	*Mrig*	42	33.18	2.78	8.38
		Hast	2	1.00	0.04	4.00
		Ambe	39	23.07	2.92	12.66
5.	926 ≥	*Mrig*	33	19.12	3.97	20.76
		Ambe	34	26.36	3.85	14.60

Varietywise Per Hectare Physical and Economic Losses of Pomegranate Orchards Affected due to Oily Spot Disease

The information on varietywise per hectare physical and economic losses of pomegranate orchards affected due to oily spot disease is given in table 25.5.

It is revealed from the table that, at the overall level, per hectare average productivity was 71 qtls which is 64 per cent less than the normal productivity and actual losses were 11 quintals i.e. in monetary returns it was Rs. 25040 (18 %). However, considering the normal productivity, losses were less by 129.20 qtls/ha (65%). Net returns were Rs.71263/ha while calculating the net returns only input costs were considered.

Baharwise analysis reveals that area under *Hast bahar* was only one per cent, while there was no significant difference between area under Mrig and *Ambe bahars*. In order to control the further dissemination of oily spot disease and to avoid the congenial climatic condition our university has recommended for taking the Hast bahar and early *Ambe bahar.*

Table 25.5: Variey wise per hectare losses of pomegranate orchards affected due to oily spot disease.

Sr. No	Variety	Bahar	Area Total	Area Affected	Affected Area %	No. of orchards	Productivity actual (q)	Net returns (Rs)	Actual Losses Physical (q)	%	Economic (Rs.)	%	Productivity normal (q)	Losses normal (q)	%	Price received (Rs./kg)
1.	Ganesh	*Mrug*	33.43	3.42	10.23	23	78.51	32341	14.24	18	23161	21	200	121.49	61	
		Hast	0.40	0.01	2.50	1	225.00	381063	3.81	2	6483	2	200			
		Ambe	14.22	1.65	11.60	19	85.65	64570	10.02	7	10484	7	200	114.35	57	14.15
	Overall		48.05	5.08	10.57	43		82.12	12.79	15	18946	15	200	117.88	59	
2.	Bhagwa	*Mrug*	31.46	3.79	12.05	36	76.35	129779	18.75	24	48742	24	200	123.65	62	
		Hast	–	–	–	0	–	–	–	–	–	–	–	–	–	–
		Ambe	36.49	4.25	11.65	44	78.53	118123	8.59	11	25316	13	200	121.47	61	
	Overall		69.95	8.04	11.49	80		77.47	13.52	17	36681	19	200	122.53	61	26.27
3.	Arakta	*Mrug*	12.33	1.13	9.16	5	33.16	-9646	2.13	6	2695	6	200	166.84	83	
		Hast	0.60	0.03	5.00	1	158.3	222500	8.33	5	15833	5	200	41.70	21	

Contd...

Sr. No	Variety	Bahar	Area		Affected Area %	No. of orchards	Produc tivity actual (q)	Net returns (Rs)	Actual Losses			%	Producti vity normal (q)	Losses normal		Price received (Rs./kg)
			Total	Affected					Physical (q)	%	Economic (Rs.)			(q)	%	
		Ambe	13.67	2.34	17.12	15	17.68	-51102	3.65	21	3217	19	200	182.32	91	–
	Overall		26.60	3.50	13.16	21		29.25	3.40	12	3655	9	200	170.75	85	14.75
4.	Mrudula	Mrug	1.30	0.23	17.69	4	33.08	-22077	0.30	1	438	1	200	166.92	83	–
		Hast	–	–	–	–	–	–	–	–	–	–	–	–	–	–
		Ambe	6.30	0.61	9.68	5	38.10	159	4.77	13	6748	12	200	161.90	81	–
	Overall		7.60	0.84	11.05	9		37.24	4.01	11	5668	10	200	162.76	81	14.20
	All varieties	Mrug	78.52	8.57	10.91	68	62.64	73.80	15.25	21	33011	23	200	126.20	63	19.93
		Hast	1.00	0.04	4.00	2	63.33	63200	2.44	4	4703	4	200	136.67	68	18.21
		Ambe	70.68	8.85	12.52	83	67.69	67.69	7.38	11	16785	12	200	132.31	66	20.62
	Total		152.20	17.46	11.62	153	71263	71.00	11	16	25040	18	200	129.20	65	20.25

The varietywise analysis shows that, minimum productivity was observed in Arakta variety (19.25 q/ha) hence losses compared to normal productivity were recorded at highest level (85%) followed by Mrudula variety (81%), Bhagwa (61%) and minimum by Ganesh (59%) variety.The baharwise losses shows that Ganesh (61%),Bhagwa (62%) and Mrudula (83%) varieties shows maximum losses in Mrug bahar while, in Arakta variety maximum losses were shown in Ambe bahar. It was also observed that losses in Ambe bahar were 56 and 61 per cent, respectively in case of Ganesh and Bhagwa variety.

In case of per kg. price received it was observed that at overall level farmers received Rs. 20.25/kg and within the three bahars maximum price was received in Ambe bahar (Rs. 20.62) followed by Mrug bahar (Rs. 19.93) and hast bahar (Rs. 18.21), it means there is no remarkable difference among the three bahars. In case of various bahars, price received per kilogram was maximum in case of Bhagwa variety (Rs.26.27/kg) and in other varieties like Ganesh, Mrudula and Arakta the price received on an average was Rs. 14/kg.

Productivity Trend of Pomegranate Orchards

The productivity trend for the last three years was calculated and is presented in table 25.6.

Table 25.6 : Productivity Trend of Pomegranate Orchards

Sr. No.	Variety	Bahar	Productivity (q/ha)		
			2004-05	2005-06	2006-07
1.	Ganesh	*Mrug*	135.58	156.94	78.51
		Hast	-	125	225
		Ambe	69.97	63.29	85.65
	Bhagwa	*Mrug*	92.09	77.57	76.35
		Hast	59.62	112.5	-
		Ambe	87.06	89.66	78.53
	Arakta	*Mrug*	88.49	87.15	33.16
		Hast	-	-	158.3
		Ambe	104.11	103.08	17.68
	Mrudula	*Mrug*	38.33	66.67	33.08
		Hast	30.00	-	-
		Ambe	94.44	112.70	38.10

It is observed from the above table that the occurrence of oily spot disease observed in all varieties and in all bahars. So there was no specific trend was observed during the last three years, but in case of Ganesh variety the productivity was increased in the year 2006-07 while, in Arakta variety it was declined tremendously.

Awareness and Adoption of Control Measures Recommended by MPKV, Rahuri to Control the Oily Spot Disease

The Awareness and adoption of control measures recommended by MPKV, Rahuri to control the oily spot disease is presented in the table 25.7.

Table 25.7: Awareness and Adoption of Control Measures Recommended by MPKV, Rahuri to Control the Oily Spot Disease

Sr. No.	Recommendation	Awareness (%)	Adoption (%)
1.	Use of unaffected planting material	25.25	0
2.	Use of Hast bahar	45.33	2.13
3.	Dipping of implements in 1% dettol	0	0
4.	Destroying the affected leaves, flowers *etc.*	98.00	38.78
5.	Spacing 4.5X3m	56.67	0
6.	Use of bleaching powder at the time of bahar	35.83	12.24
7.	Use of 10% bordo paste immediately after cutting	98.00	68.27
8.	Use of 5% urea f or shading of affected flowers leaves *etc.*	0	0
9.	Cutting of affected plant below 2 inch level	78.33	58.59
10.	Use of 1 % bordo mixture after first spraying	99.10	95.10
11.	Use of Streptocycline + copper oxoide + sticker at the time of second spraying	78.56	65.13
12.	Use of 4% bordomixture at the time of third spraying	0	0
13.	Use of Streptocycline + carbendenzim at the time of fourth spraying	45.33	35.61
14.	Use of Streptocycline at the time of bad weather	46.45	42.17
15.	Use of 1% bordomixture and giving 3 to 4 months rest to the plants	50.12	48.15
16.	Cleaning campaign at village level once in a year	40.18	1.19

The pomegranate growers were not aware about the control measures like dipping of implements in 1% dettol, use of 5% urea for shading of affected flowers leaves *etc.* and use of 4% bordomixture at the time of third spraying. Majority of the growers aware about destroying the affected leaves, flowers *etc.*, use of 10% bordo paste immediately after cutting, use of 1 % bordo mixture after first spraying Cutting of affected plant below 2 inch level and use of Streptocycline + copper oxoide + sticker at the time of second spraying and the they have adopted these control measures. In case of the control measure like use of Hast bahar and cleaning campaign at village level once in a year, the pomegranate growers were aware but they have not adopted.

Conclusions

1. The area under pomegranate on sample farms was 1.31 ha., of which the proportion area under different varieties of pomegranate viz; Ganesh, Bhagwa, Mrudula and Arakta was 37.40, 47.33, 3.05 and 11.45 per cent, respectively.

2. The proportion of area under pomegranate affected by the oily spot disease was 10.69 per cent on the sample farms.

3. Among the four varieties of pomegranate, the maximum area under Arakta variety was affected followed by Ganesh, Bhagwa and Mrudula varieties.

4. Among the three bahars, maximum area was affected in Ambe bahar followed by Mrug and Hast bahars.

5. The maximum area affected in the case of plant density above 926 in mrug bahar.

6. The per hectare average productivity on sample farms was 71 quintals which is 64 per cent less than the normal productivity and actual physical losses were 11 quintals and the economic were Rs. 25040.00.

7. The baharwise and varietywise physical and economic losses were maximum in mrug bahar of Ganesh, Bhagwa and Mrudula varieties and Ambe bahar in the case of Arakta variety.

8. The maximum price was received to the fruits of Ambe bahar (Rs. 20.62/kg) followed by Mrug bahar (Rs. 19.93/kg) and Hast bahar (Rs. 18.21/kg).

9. The highest per kg. Price of Rs. 26.27 was received for the Bhagwa variety and Rs. 14 to 15 for Ganesh, Mrudula and Arakta varieties.

10. The pomegranate growers are not aware of the control measures like dipping of implements in 1% dettol, use of 5% urea spray on affected flowers leaves *etc.* and use of 4% bordomixture at the time of third spraying.

11. Majority of the growers aware about destroying the affected leaves, flowers *etc.*, use of 10% Bordeaux paste immediately after cutting, use of 1 % Bordeaux

mixture after first spraying Cutting of affected plant below 2 inch level and use of Streptocycline + copper oxoide + sticker at the time of second spraying and the they have adopted these control measures.

Suggestion

The occurrence of oily spot disease was more in the case of Arakta variety in Ambe bahar so that the economic losses were to the tune of 18 per cent. The pomegranate growers are not fully aware of and adopting partially the control measures recommended by the university. Hence, there is a need to avoid Ambe bahar for Arakta variety, adoption of proper planting distance and control measures recommended by the university for minimizing the economic losses due to oily spot disease.

26

Trends and Perspectives of using Biopesticides for Enhanced Tolerance Against Diseases in Pomegranate

Girigowda Manjunath[1] and Jyotsana Sharma[2]

[1]Biocontrol Laboratory, University of Horticultural Sciences, Bagalkot-587 104, Karnataka
[2]ICAR-National Reseach Cente on Pomegranate, Solapur-413 255, Maharashtra
E-mail: Gmanjunath2007@gmail.com

Introduction

Pomegranate bacterial blight and wilt have evolved as epidemics in the recent times with high preferential management options. Whilst, other diseases caused by species of *Alternaria, Colletotrichum, Cercospora etc.,* are also having a serious impact on fruit production of pomegranate. These diseases are chiefly managed by using synthetic pesticides of high cost and few of them have residual toxicitywith possible chances of causing resistance in targeted pathogen, hence, the reduction of synthetic molecules use in pomegranate is increasingly demanding. Further, most of synthetic molecules are counterproductive, are increasing input cost and are unsafe to health and environment, therefore, use of synthetic molecules has to be critically evaluatedand it is desirable to replace/integrate chemical control methods with less toxic biological methods.

The use of bioagent formulations comprising *Pseudomonos fluorescens* and *Bacillus* spp. has been a component of integrated disease management and their effect on reducing the impact of disease was appreciable,albeit, the scientific publications on their use are very scanty except academic interest work in the form of dissertations and part of the thesis (Yenjerrappa *et al.*, 2013). The inhibitory actions of *Pseudomonas* spp. was reported against anthracnose disease caused by *Colletotrichum* spp. and it is expected to antagonize other foliar pathogens including *Alternaria* spp. as well (Satreddy, 2009). Mycorrhizae associated actinomycete *Streptomyces canus* as a bio-inoculant was found promising for improvement of growth promotion and for improving tolerance against bacterial blight (Poovarasan *et al.*, 2013). The species of *Trichoderma* like *hamatum, harzianum, koningii* were found effective *in vitro* against Xap (Apet *et al.*, 2013), however, these are usually soil inhabitants and their efficacy on field conditions at phylloplane has to be evaluated for exploring into commercial purpose.

The biocontrol agents (BCA) such as *P. fluorescens, Bacillus* spp. and various fungal antagonists werefound beneficial, as most of them are known to release growth promoting agents like gibberellic acid, IAA, cytokines, auxins, lytic enzymes *etc.*, in plants contributing to improve the biomass, root proliferation, vigour of plants and overall tolerance against biotic and abiotic stresses (Indi, 2010). Once BCA is introduced and favourable conditions are created for the antagonist to establish into thresholds levels, the BCA continually keeps the pathogen populations under controlwith high degree of host specificity.

Further, use of bacteriophages to manage bacterial blight of pomegranate was also reported as a pathogen targeted approach unlike beneficial microbes (Kiran Kumar 2007; Yenjeerappa *et al.*, 2009), however, development of bacteriophage formulations and their application under field conditions is challenging. In addition, copper compounds and their residues known to have deleterious effect over the phages multiplication, so it demands compatibility assays with other agrochemicals (Dennehy, 2011).

Biopesticides Products Positioning for Pomegranate

A survey was undertaken on recording the use of biopesticides for pomegranate cropping in Karnataka. A range of bacterial and fungal BCAs with other plant based biologicals have been under use in pomegranate (Table 26.1). The purpose of this paper is also to provide the information on use of different biopesticides which is very scattered, and analyzing the practical feasibilities and relevance of their use in continuum. The formulations containing bacteria and fungi primarily used by the farmers were recorded as microbial pesticides and formulations based on plants and seaweed were recorded as botanical pesticides. The interaction with farmers on efficacy of these products revealed that these formulations on application are providing inconsistent results at field levels and bioactivity of these products is a matter of concern and proving to be a major bottleneck for suggesting to a large scale field application. Further, quantification of this product impact is fairly difficult however; few of the formulations were reported to improve the vigour and quality of fruits that needs further data validation. In addition, BCA products being marketed currently to pomegranate farmers are not correctly labeled about optimized efficacy, stability of the product, safety, storage and biological fitness being capable of replicating in the applied field.

Identification and Evaluation of Antagonists

A series of antagonists were isolated from native fields of pomegranate and their efficacy was evaluated against bacterial blight pathogen and wilt fungus both at Bagalkot and ICAR-National Research Centre on Pomegranate, Solapur in the project mode. The obtained data is presented in this section.

Table 26.1: Biological Formulations Currently in use at Various Fields of Pomegranate for Management of Bacterial Blight and Increase in Plant Vigour

Formulations	Organism associated	Manufacturer	Remarks/claims
Probiotic Growbax	An effective combinations of naturally occurring beneficial microbes	Microbox India limited, Hyderabad	Enhances resistance against pests and pathogens.
Probiotic Sulfex	*Bacillus* spp.	Microbox India limited, Hyderabad	For enhance immunity
Sudo	*Pseudomonas fluorescens*	Kan Biosys Pvt. Ltd.	Effective against a wide variety of seed and soil borne plant pathogenic fungi
Nemagon	*Paecilomyces lilacinus*	Green life Biotech Laboratory	To attack nematodes and their eggs without being harmful to any plant species
Xoton Plant immunizer	-	West cost Rasayan International Pvt. Ltd.	Increases immunity of plant to fight against phycometes fungti like downy mildew, late blight, quick wilt and root rot. *Etc.*
Top life Super smart -	See weed extract	ROOTS-Florentine Curtis. Pte. Ltd. Singapore and Nashik Marketed by; TL Agro. Pvt. Ltd. Toplife House, Pune, Maharastra	For increased immunity

Contd...

Formulations	Organism associated	Manufacturer	Remarks/claims
Phytoalexin 84	Herbal extracts	Akshay Chemicals, Ratnagiri	For increased immunity
Kil Tel	Chitosan based	Camlin Fine Sciences Ltd., Mumbai	Improving plant growth and defense mechanism
Zanthonil	*Oscimum sanctum*, ginger oil, clove oil	Sanardhini Agro. Pvt. Ltd., Satara	Defense booster
Kalisena SA	*Aspergillus niger*	Cadila Pharmaceuticals, Ahmadebad	Induction of resistance, control of soil borne wilt pathogens, improved growth and production
Josh Super	Vesicular Arbuscular Mycorrhizae	Cadila Pharmaceuticals, Ahmadebad	Better rooting and enhanced growth

Antagonist Evaluation of Bacterial Blight Pathogen

The efficacy of *Pseudomonas fluorescens* (Pf-48) was evaluated under greenhouse conditions. A total of 6-sprays of *P. fluorescens* were used in the time gap of 60 days with 5ml and 10ml/l concentration containing CFU 2×10^8/ml under greenhouse condition. An average disease incidence of 54.67 percent reduced to 11.49 percent after 60 days of treatment at 5ml/l concentration. A similar reduction was also recorded at 10 ml/l with no significant difference among 5 and 10ml/l. Whereas, in untreated control plants, disease incidence increased from 54-79 percent. Interestingly apart from the reduction in total disease incidence significant increase in healthy leaves and growth promotion was recorded in treated plants compared to untreated plants (Fig. 26.1).

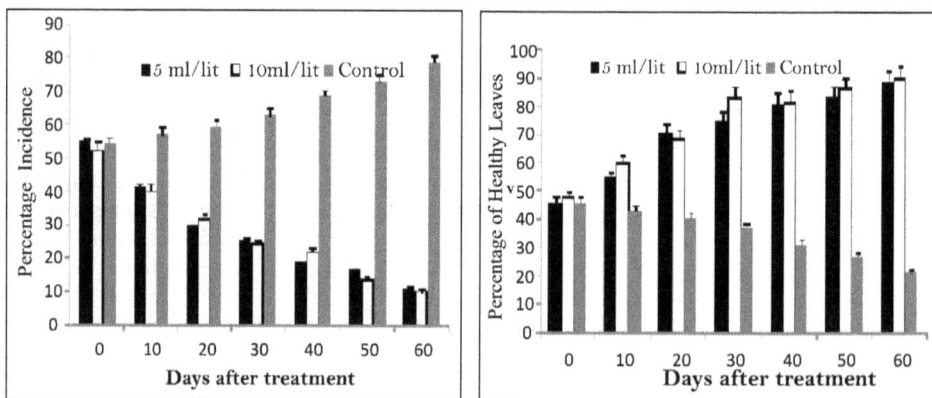

Fig. 26.1: Efficacy of *Psedomonas fluorescens* against Bacterial Blight of Pomegranate with Percent Disease Incidence (Left Panel) and Enhancement of Healthy Leaves (Right Panel)

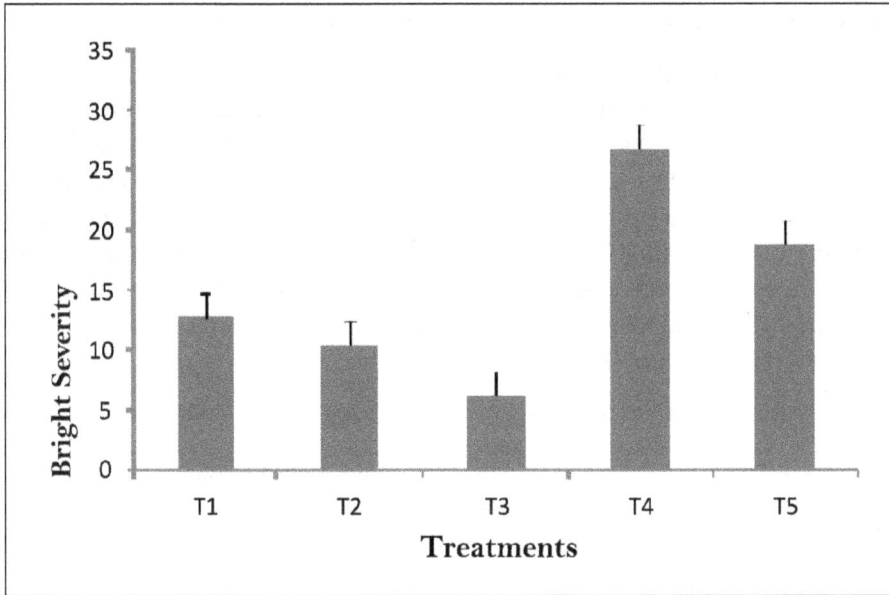

Fig. 26.2: Effect of Different Treatment Schedules with *Psedumonas fluorescence* (Pf48) Formulations on Bacterial Blight Severity Under Field Conditions in *hasta bahar* Conditions at Bagalkot

Treatment Details

T1   1 spray of copper oxychloride (3g/l) +streptomycin sulphate (0.5g/l)

T2   1 spray of Pf 8day after copper oxychloride (3g/l) + streptomycin sulphate (0.5g/l) 2 spray in 8 days interval after Copper oxychloride (3g/lt)+ streptomycin sulphate

T3   (0.5g/l)

T4   Pf 2 Sprays 8 days interval

T5   2 sprays of copper oxychloride (3g/l) + streptomycin sulphate (0.5g/l)

Pseduomons fluorescens (Pf48) based talc formulation was applied in different schedules as detailed in the fig. 26.2 with different combinations of protectants like copper oxychloride and antibiotic- streptomycin sulphate. The maximum reduction of disease was recorded when Pf applied was after eight days of copper oxychloride and antibiotics application. Interestingly, less severity was seen in treatment three than treatment five, where two applications of copper oxychloride and antibiotic were imposed.

Field Evaluation of Bacterialantagonists at Solapur

Another Field trial data on using *Pseudomonas fluorescens* andcommercially available formulation Bacterimax against bacterial blight disease is also presented here. The experiment was conducted in farmer's field at Solapur, Maharashtra in *hasta bahar* season in variety Bhagwa. Fruits harvested in May 2008 were observed for bacterial blight infection. *P. fluorescens* isolate - 2 from NCIPM, New Delhi and 1 commercial formulationBacterimax (Table 26.2), significantly reduced bacterial blight incidence over untreated control and were at par with antibiotic streptocycline (streptomycin sulphate 90% + oxytetracycline. However, it is interesting to note that all the tested isolates reduced the severity considerably.

Table 26.2: Efficiency of Bioagents in Checking Bacterial Blight in Field

Molecules tested	Sensitivity reactions
2-Bromo-2 nitro propane 1-3 diol	+
Streptomycin sulphate+ tetracycline hydrochloride (9:1)	+
Copper oxy chloride	+
Calcium	-
Magnesium	-
Boron	-
Gibberellic acid	Inhibition at 750ppm and above
Tilt	Inhibition at 500 ppm and above
Carbendazim	-
Ethanol	-
Ethrel	-
Sodium hypochlorite	+

Antagonist Biocontrol Studies on Wilt Fungus

A total of 260 isolates of *Trichoderma* spp. were recovered from rhizosphere soils of pomegranate field. The recovered species were subjected for their antagonistic efficacy against *Ceratocystis fimbriata*, a wilt causing pathogen. The data presented in the table 26.4, listed the effective isolates with more than 88 percent of inhibition over the control. These effective isolates were taken into greenhouse (pot culture studies) after these were formulated with talc powder amended with carboxy methyl cellulose.

Table 26.4: Data on *In vitro* Antagonistic Assay Using
Trichoderma spp.* against *C.fimbriata

List of effective isolates	Inhibition of C. fimbriata over the control (%)
Pt5	100.00±2.8
Pt110	091.42±3.2
Pt112	091.42±1.9
Pt13	097.14±2.5
Pt117	088.57±3.8
Pt24	094.28±1.5
Pt43	100.00±2.1
Pt48	100.00±3.2
Pt251	094.28±1.7
Pt255	097.14±1.9

A total of 150 different *P. fluorescens* were recovered from both phylloplane and rhizoshperic soil of pomegranate growing area. The isolated agents were subjected for their efficacy against *C. fimbriata* and the isolates showing promising results arelisted in table 26.4. The effective isolates were taken into pot culture studies along with effective isolates of *Trichoderma* sp.

Table 26.5: Data on *In vitro* Antagonistic Assay Using *P. fluorescens*
against *C. fimbriata*

List of effective isolates	Inhibition of C.fimbriata (%)
PPf 102	92.85±3.4
PPf 12	54.28±2.8
PPf 34	94.28±3.1
PPf 25	57.14±2.6
PPf 26	90.00±3.1
PPf 36	71.42±2.8
PPf 143	92.85±3.2
PPf 48	95.71±2.1
PPf 51	94.28±3.4
PPf 155	64.28±2.6
PPf 46	64.28±2.9

Conclusion

The use of biopesticides assumes significance under the ambit of IDM enabling the effective management of diseases with reducing load of toxic pesticides and their residues in pomegranate. However, if we compare the status of biopesticides use *vis-a-vis* with other cropping situations, the picture is not satisfactory. A regulation of biopesticides quality is a matter of high concern to ensure the better BCAs with longer shelf life and appropriate delivery methods for their continuous use in pomegranate. The re-isolation of applied antagonists from pomegranate foliage has a little chances of recovery when it is used through talc formulations, therefore, emulsifier based formulations are needed for sustaining activity of BCAs temporally with long persistence. Further, biopesticides products efficacy has to be assessed periodically once releasedfor use byfarmers. The effective compatible consortium of bioagents and nutrients mobilizing inoculants are necessary to popularize the biopesticides formulations for pomegranate. Further, greater cooperation with industrial scientists required for transferring the identified technology into mass-scaling for serving the products in large scale.

27

In vitro Evaluation of Botanicals Against *Xanthomonas axonopodis* pv. *punicae* Causing Bacterial Blight of Pomegranate

C. Gopalakrishnan

Division of Plant Pathology, ICAR-Indian Institute of Horticultural Research,
Bengaluru – 560 089, Karnataka
E- mail: gopkran@iihr.ernet.in

Introduction

Pomegranate (*Punica granatum* L.) cultivation has emerged as remunerative enterprise in the states of Maharashtra, Karnataka, Gujarat and Andhra Pradesh. Consequently, there is a steady increase in area and production of the crop and export of fruits. However, bacterial blight disease caused by a bacterium, *Xanthomonas axonopodis* pv. *punicae* (Hingorani and Singh 1959) Vauterin, Hoste, Kersters & Swings 1995, has become cause of concern, as many farmers have lost their crops as well as plantations due to this disease. Pomegranate growers use bactericides like Bordeaux mixture, copper oxychloride, Streptocycline and Bronopol to control the disease but end up with inadequate control. The natural plant products derived from plant species has the capacity to control diseases caused by viruses, bacteria and fungal pathogens. Research focused on plant-derived natural bactericides and their possible applications in agriculture to control plant bacterial diseases has intensified as this approach has enormous potential to inspire and influence modern agro-chemical research. Hence, the present study is focused on *in vitro* evaluation of various plant extracts and oils against *X. axonopodis* pv. *punicae* causing bacterial blight in pomegranate.

Materials and Methods

The experiment was designed to evaluate the antibacterial activity of various botanicals against *X. axonopodis* pv. *punicae* under *in vitro*. The leaf extracts of 19 plant species and three plant essential oils (Table 1) were screened for their antibacterial activity against the bacterial blight pathogen, *X. axonopodis* pv. *punicae* using agar well diffusion method (Mahajan *et. al.*, 1991). The inoculum of *X. axonopodis* pv.*punicae* was prepared from the pure culture of the bacterium obtained from freshly infected pomegranate leaves. The bacterium was isolated using LB agar medium. The inoculum was prepared by suspending the bacterial growth from 523-mediumin sterile distilled water and the concentration of the inoculum was adjusted to 0.3 OD at 600nm wave length (1.0×10^8cfu/ml) using spectrophotometer.

Fresh extracts from the leaf, bulb (garlic and onion), rhizome (ginger) and fine powder (turmeric and asafoetida) were used for carrying out the experiment (Table 27.1). One g of the leaf, bulb and rhizome samples were thoroughly washed with tap water and then rinsed in sterile distilled water. The samples were ground in mortar and pestle with 10.0 ml sterile distilled water separately. In the case of turmeric and asafoetida, one gram of power was dissolved in 10.0 filtrate was used as 10% plant extract. The clear extract was used for testing its antibacterial activity against *X. axonopodis* pv. *punicae* under *in vitro*. The commercial essential oils, viz., *Azadirachta indica* A. Juss., *Pongamia glabra* L. and *Syzygium aromaticum* (L.) Merril & Perry were diluted to 10% using the solvent Di-Ethyl ether before loading into the wells.

The agar medium(523medium) was prepared in sterile Petri plates (90 x 15 mm) and 100 µl of bacterial inoculum (1.0×10^8 cfu/ml) was evenly spread ontothe medium. 5.0 mm diameter well was made in each agar plateusing sterilized cork borer. 50 µl of botanical extracts and essential oilswere loaded into the wells in Petri plates separately. The commercial bactericide, K-Cycline (Streptomycin sulphate 90% w/v and Tetracycline hydrochloride 10% w/v) 500 ppm and Bronopol (2-Bromo -2-nitro propane -1,3 diol) 500 ppm were used as standards for comparison and sterile distilled water was used as control check. The whole experiment was carried out under aseptic condition inside the sterile laminar flow bench. The Petri plates were left undisturbed to allow diffusion of the sample for 2-3 h and incubated at 30°C temperature for 48-72 h. Observations were recorded periodically on inhibitory zones(diameter in mm) produced by plant extracts and essential oils against *X. axonopodis pv.punicae*around the wells.

Results and Discussion

The results of the antibacterial activity of 19 leaf extracts and three plant essential oils against *R. solanacearum* is shown in Table 31.1 as zone of inhibition (diameter in mm). It was observed that of the 22 botanicals tested, three extracts from, *A. sativum*, *L. inermis* and *P. betle* and one plant essential oil, *S. aromaticum* showed antibacterial activity against *X. axonopodis* pv. *punicae*. The maximum inhibition zone of 60.0 mm diameter was observed in *S. aromaticum* treatment followed by *A. sativum*, which showed inhibition zone of 30.0 mm diameter. Whereas, *P. betle* and *L. inermis* showed inhibition zone of 20.0 mm diameter against *X. axonopodis* pv. *punicae*. However, the other plant extracts and plant essential oils tested did not show any antibacterial activity against *X. axonopodis* pv. *punicae*. The commercial bactericides, Bronopol and K-Cycline used as standard check showed inhibition zone measuring 40.0 and 32.0 mm diameter, respectively, and water used as control check did not show any inhibition against *X. axonopodis* pv. *punicae*. Results also showed that *S. aromaticum* (60.0 mm diameter) and *A. sativum* (30.0 mm diameter) showed significantly highest inhibition as compared to the bactericides, Bronopol (40.0 mm diameter) and K-Cycline (32.0 mm diameter) tested against *X. axonopodis* pv. *punicae*. However, the antibacterial activity of *L. inermis* and *P. betle* (20.0 mm diameter) was significantly low as compared to *S. aromaticum* and *A. sativum* (Table 27.1).

In the present study, *P. betle* exhibited antibacterial activity against *X. axonopodis* pv. *punicae*. Pelczar *et. al.* (1993) and Pauli (2002) suggested the antimicrobial property of *P. betle* extract is due to its various chemical components such as the sterol, hydroxyl chavicol, eugenol and phenolic compounds (chavicol). Moreover, other chemical components such as fatty acid (stearic acid and palmitic acid) and hydroxyl fatty acids esters (hydroxyl esters of stearic, palmitic and myristic acids) exist in betel plant are also known to possess antibacterial properties (Liao *et. al.,* 1999; Bhattacharya *et. al.,* 2007). Betel leaf stalk extract was also found to have antibacterial activity against Gram positive and Gram negative bacteria (Pompee *et. al.,* 2013). Hence, *P. betle* also has high potential to be used as bactericide against bacterial blight of pomegranate.

Though garlic extract and clove oil have shown to have high potential against several microorganisms (Huang and Lakshman, 2010; Jeyaseelan *et. al.,* 2010) including *X. axonopodis* pv. *punicae* in the present study, the commercial exploitation in the management of bacterial wilt may be very expensive. Since, *Piper betle* (Betel) and *L. inermis* (Henna) are cheap commodities and will be most economical and potential alternative to common antibiotics, if their potential is exploited in the management of bacterial blight of pomegranate.

Table 27.1: Antibacterial Effect of Botanicals Against
Xanthomonas axonopodis pv. *punicae*

Sl. No.	Botanical Name	Common Name	Name of family	*Mean Zone of inhibition (mm, diameter)
1.	*Allium cepa* L.	Onion	Amaryllidaceae	0.00
2.	*Allium sativum* L.	Garlic	Amaryllidaceae	30.00
3.	*Azadirachta indica* A. Juss. (leaf)	Neem	Meliaceae	0.00
4.	*A. indica* A. Juss. (oil)	Neem	Meliaceae	0.00
5.	*Calotrophis gigantea* R.Br..	Milk weed	Apocynaceae	0.00
6.	*Catharanthus roseus* (L). G. Don	Periwinkle	Apocynaceae	0.00
7.	*Coleus forskohlii* Brig.	Coleus	Lamiaceae	0.00
8.	*Curcuma longa* L.	Turmeric	Zingiberaceae	0.00
9.	*Datura Stramonium* L.	Datura	Solanaceae	0.00
10.	*Ferula assa-foetida* L.	Asafetida	Apiaceae	0.00
11.	*Lawsonia inermis* L.	Henna	Lythraceae	20.00
12.	*Leucas aspera* Willd. (L.)	Thumbai	Lamiaceae	0.00
13.	*Mentha longifolia* (L.) Huds.	Mentha	Lamiaceae	0.00
14.	*Nerium oleander* L.	Oleander	Apocynaceae	0.00
15.	*Ocimum sanctum* L.	Tulsi	Lamiaceae	0.00
16.	*Piper betle* L.	Betel	Piperaceae	20.00
17.	*Pongamia glabra* L.	Pongamia	Fabaceae	0.00
18.	*Syzygium aromaticum* (L.) Merril & Perry	Clove	Myrtaceae	60.00
19.	*Tinospora cordifolia* (Thunb.) Miers	Guduchi	Menispermaceae	0.00
20.	*Vitex negundo* L.	Chaste tree	Verbenaceae	0.00
21.	*Withania somnifera* (L.) Dunal	Ashwagandha	Solanaceae	0.00
22.	*Zingiber officinale* Roscoe	Ginger	Zingiberaceae	0.00

Contd...

Sl. No.	Botanical Name	Common Name	Name of family	*Mean Zone of inhibition (mm, diameter)
23.	K-Cycline (500 ppm)	---	---	32.00
24.	Bronopol (200 ppm)	---	---	40.00
25	Control (water)	---	---	0.00
	CD (P 0.05)			0.06
	S.Em±			0.02
	CV%)			4.69

*Mean of three replications

28

Experiences of BBD Management in Pomegranate: A Case Study

L.R. Tambade and P.A. Gonjari

Krishi Vigyan Kendra, Solapur-413 255, Maharashtra

Introduction

Pomegranate (*Punica granatum*) is grown in tropical and subtropical regions of the world. The total area under cultivation of pomegranate in India is 193.00 thousand ha and production is around 220 thousands tons. Maharashtra is the leading producer of Pomegranate followed by Karnataka, Gujarat and Andhra Pradesh. Ganesh, Bhagwa, Ruby, Arakta and Mrudula are the different varieties of pomegranate produced in Maharashtra. In india pomegranate is commercially cultivated in Solapur, Sangli, Nasik, Ahmednagar, Pune, Dhule, Aurangabad, Satara, Osmanabad and Latur districts of Maharashtra. Bijapur, Belgaum and Bagalkot districts of Karnataka and to a smaller extent in Gujarat, Andhra Pradesh and Tamil Nadu.

Brief Background

The area under fruit crops in Maharashtra state has reached higher level due to efforts made by State Government through employment guarantee scheme (EGS) 1999. Now a day; Pomegranate grown on more than 1,28,000 ha. area in Maharashtra, contributing 37,746 ha. area by the Solapur district. The fruit pomegranate has given enormous level of status to Pomegranate growers in terms of social, economics in arid zones having light soils.

But, since last 8-9 years the Pomegranate has badly affected by Bacterial Blight disease on epidemic level. The whole orchards of number of villages in Solapur district has been infected and farmer loosen every profit from it. KVK, Solapur demonstrated the BBD management protocol in KVK operational area specially in Papari cluster during 2008-09 to 2011-12 very precisely to educate the pomegranate growers regarding management of BBD through different extension tools.

Mr. Vitthal Annasaheb Karande a youth farmer from village Wadachiwadi, Tal: Mohol, Dist: Solapur (MS) having 15 acres of irrigated land. He cultivating only Pomegranate crop on 09 acres area out of which 6 acres pomegranate orchard is 5 years old while remaining 2 acres is 2 years old orchard. He is only concentrating on pomegranate crop specially on Bacterial disease management. Remaining 6 acres of land is under grape (3 acre), onion (2acre) and one acre under farm pond and farm house.

Mr. Karande, with his efforts and technical back up by KVK, Solapur he secured 39 tons of Pomegranate fruits from 2200 plants (6 acre area) and got Rs. 34,22,009/- (Thirty Four lacs Twenty two thousand and nine rupees) as gross return in Mrig Bahar, 2010 without exporting his produce to other countries.

Mr. Vitthal Karande from Wadachiwadi, Tal: Mohol, Dist: Solapur has emerged as lesson giving farmer to other Pomegranate growers by assuring highest tonnage and net profit from 6 acres of Pomegranate orchard.

Mr. Karande is having Pomegranate of 06 acres planted in June, 2005. The planting distance is 12 x 8 ft. includes 2200 plants in 6 acres. Remaining 3 acres of pomegranate was planted during June 2009. He has taken mrig bahar in 2006 and 2007 but due to heavy incidence of Bacterial Blight Disease the bahar failed during that season. During 2007-08 he got Rs. 90,000/- and 1,40,000 per acre as net profit from Pomegranate. As increase in net profit, he got interest in Pomegranate farming at same time he came in contact with KVK scientists.KVK, Solapur conducted topical PRA in village papri and taken Wadachiwadi as Satalite village during December, 2007 and adopted village for solving the problems specially in Pomegranate, Ber & Banana crops and dairy enterprise.Krishi Vigyan Kendra, Solapur had conducted FLD on Bacterial Blight management and also on integrated nutrient management in Pomegranate at village Papari and Wadachiwadi Tal: Mohol in Ambe Bahar 2008, 2009 and Mrig 2010. Mr. Vitthal Karande was one of the beneficiary farmer for FLD on BBD management conducted by KVK, Solapur.

He is also one of the beneficiaries of different KVK activities. Training on BBD management, Method Demonstrations, Video shown on BBD & INM in Pomegranate, group discussion for GMP (good management practices) adopted in Pomegranate, frequent visits of KVK, scientists and tele-helpline service given by KVK. Mr. Karande develop good rapport with KVK scientist.

Now the path for record yield and income was going to establish with the joint efforts of KVK and Mr.Karande. After the thorough discussion with KVK scientist Mr. Vitthal Karande decided to take Mrig Bahar in 2010. The orchard managment planning has been decided. The Water stress for orchard started from end of April. The whole work of leaf fall (with etheral @ 2 ml / lit.) and prunning was done in May, 2010. The most critical deciding factor was, when to give first irrigation to stop water stress of orchard. During last week of May 2010 first irrigation was given. For 2010, it was became crucial factor, as orchard was escaped from heavy rains and fruit setting was completed before 20th June, 2010.

But due to heavy rains from June 2010 which crosses the average rainfall limit of Solapur district had made the farmer worried about BBD problem. But Mr. Karande comes out from this problem by adopting the recent technologies developed by MPKV, Rahuri and NRCP, Solapur for the management of BBD in Pomegranate which was transferred through KVK scientist.

The light prunning in March, 2010 and sufficient application of FYM, Neem Cake and other organic matter after last bahar fruit harvest contributed more C:N ratio and food storage in the stems which has given more no. of flower (female + hermaphrodite) and fruit setting in from July 2010 onwards.

Mr. Karande did the total expenditure of Rs. 6,02,991/- on improved cultivation of Pomegranate on 6 acres orchard as per the guidelines of KVK, Scientist.

Record Yield

Mr. Vitthal Karande secured 39 tonnes of Pomegranate from 2200 plants (6 acres) in Mrig Bahar-2010. He got average fruit weight 200-700 g and 150-200 fruits per plant.

The Details of the Marketing of Fruits done by Mr. Karande

Sale of fruits	Market	Price secured (Rs.)	Total Income (Rs.)
24 Ton	Kolkata & Pune	87.5 / kg	21,00,000.00
10 Ton	Kolkata & Pune	120 / kg	12,00,000.00
05 Ton	Kolkata & Pune	145 / kg	7,25,000.00
		Total	40,25,000.00
Total Production Income		Rs	40,25,000.00
(-) Total expenditure in Bahar		Rs	6,02,991.00
Total Net Profit		Rs	34,22,009.00

He secured Rs. 34,22,009.00 (Rupees Thirty four lakhs twenty two thousand and nine only) as net profit from 6 acres Pomegranate orchard in Mrig Bahar, 2010.

Experiences Shared by Mr. Vitthal Karande

- The intensive care should be taken after last years fruit harvest. Most of farmers neglect this fact.

- The light prunning during March, 2010 was found helpful to increase in C:N ratio and food storage in stem.

- Group approach for orchard sanitation for management of Bacterial Blight in Pomegranate is essential and the growers are doing the same.

- Bactronol + COC combination during first two months are very effective in managing BBD.

- Like wise 0.5% Bordeaux spray when prepared at field level and spraying of KMS +H_3PO_4 are also found effective in managing BBD and other diseases.

- Application of 6BA 1gm / 100lit @ leaf emergence after pruning and root pruning are also very essential.

- If leaves and flowers are affected with BBD, Leaf shadding and according application of Bleaching powder on soil can also check BBD infestation.

Experiences of KVK

- The light to medium pruning was found beneficial for better fruit setting and its distribution throughout the plant.

- The use of COC + Streptocycline + Bactronol was found effective for second month spray.

- It was observed that due to use of INM precisely with BBD protocol the management of BBD was possible with improvement of yield up to 36.07% (Due to STBF 18.42 and BBD protocol 17.65% respectively).

- The fruit weight improves from 291-04 g to 319.7 g in Bhawa.

- Due to use of BBD protocol (27 sprays) the disease free fruits were 76.87% , where as in local it was 60.37% with 34 sprays.

- It was observed that curing of FYM with micro-nutrient enhance the growth of plant.

- Due to STBF improves fruit weight 9.87% and reduced incidence of BBD up to 8-10%.

- The BBD incidence occurred more in light soil orchard (Depth less than 20 cm) than medium soil.

Dr. S. Ayyappan, Secretary, *DARE & DG, ICAR* felicitated Mr. Vitthal Karande in presence of *DDG (Horticulture), ZPD (Zone-V) & DOR, MPKV, Rahuri* during his visits to FLD plot on BBD management in pomegranate (A/P: Wadachiwadi, Tal: Mohol, Dist: Solapur) on 26th September, 2010 on his achievements in Pomegranate cultivation and management of BBD with the technical backing of KVK, Solapur. The Director General had appreciated the efforts of Mr. Karande & outreach of KVK for solving the farmers problems.

In addition to this Dr. Omprakash, Consultant, Planning commission under NHM has visited FLD, BBD management plot along with SAO, Solapur and other dignitaries and appreciated the work done by Mr. Karande and technical backup guidance by scientists of KVK, Solapur.

Farmer Feelings About KVK

KVK, Solapur has major contribution in giving valuable guidance in time regarding the complete package of practices and effective measures for managing BBD in pomegranate by regular visit of KVK scientist as an when required. Now all the Pomegranate growers of my village coming together, and taking group measures for managing BBD in pomegranate with the help of KVK, scientists.

29

Pomegranate Wilt and its Management

Jyotsana Sharma

ICAR-National Reseach Cente on Pomegranate, Solapur-413 255, Maharashtra

Pomegranate is an important fruit crop in India. Due to high returns from this crop, it is being adopted by farmers at a fast rate.During last 5-6 years there is an increase of 80.37% in pomegranate area, 195.82% in production, 63.97% in productivity and 66.38% in export at national level. Consequently a crop once known to be hardy has become prone to various diseases and wilt/decline is the second most important disease in recent times in pomegranate. Apart fromall pomegranate growing areas in India it is reported from Iran, China and Greece.Losses in individual orchards have been recorded up to 35%.*Ceratocystis fimbriata* and *Meloidogyne incognita are the* major cause of wilt in pomegranate and species of *Xyleborus Fusarium, Rhizoctonia, Sclerotium, Macrophomina, Phytophthora* are occasionally associated.

Symptoms

As soon as yellowing/drooping/drying of leaves in some of the branches or entire plant is observed one should first diagnose the cause and then treat the plants accordingly. For this observe the roots and split open the roots and lower portion of the stems, if you observe:

- brown/gray/black discolouration of wood it is a fungus *Ceratocytis fimbriata*
- only xylem is brown it is *Fusarium* sp.

- pin holes are observed it is shot hole borer
- knots are observed on fine roots in early stage and other roots in advance stageit is nematode infestation.
- ifstem above soil level shows collar rot it can be *Rhizoctonia and/or Phytophthora*. In this case sudden toppling down of green plant can also be seen.
- If root are slimy to touch or white-black fungal growth is observed feeder roots are missing then it could be root rot by *Sclerotium or Macrophomina*

Predisposing Factors: The wilt diseases are generally aggravated due to biotic stresses particularly drought as well as excessive rain, boron deficiency in soil result in increased severity to *C. fimbriata*. Wounds natural or due to insect/nematode or human activity like pruning and inter cultural operations predispose the plants to severe infections, as the pathogen is more devastating in overindulged orchards rather than in orchards with little human activity. Stress due to flowering and fruit bearing trees also results in sudden death of the entire plant. Nematode wilt is more common in sandy soils. All commercial cultivars are susceptible to wilt and it can attack plants of all ages.

Survival: It survives as mycelia within host or as aleurioconidia in the plant host or debris in the soil. Su rvival of the *C. fimbriata* in infected pomegranate plant parts was recorded upto 2 years though survival of *C. fimbriata* up to 5 years is also reported.

Wilted Plants due to *Ceratocystis fimbriata*

Brown Dicolouration of Vascular Tissue and Wood due to   ***Fusarium wilt***
C. fimbriata

   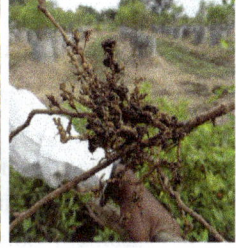

Root rot   **Collar rot**   **Shot hole borer**   **Root knot**
                                            with larva and pupa

Spread: Planting material is an important mode of primary spread to new localities. The soil used for raising the nursery planting material is an important agent for the spread of all wilt pathogens.

Spread to adjacent treesmay be through root grafts in close planting. Flood irrigation/rain water also results in similar spread. It may also infect through wounds in the roots made by insects and rodents. Pruning wounds are common entry points for *C. fimbriata*, and the fungus can be carried on pruning tools.

Management

1. **The planting material** (sapling as well as soil in which it is planted) should be free from all wilt causing agents-the fungi, insects and nematodes; use solarized/sterilized soil for planting saplings. It is advisable to take cuttings/ air layers from disease free orchards and make your own saplings using sterilized soil.

2. The **soil used for potting mixtures or soil of beds for planting new orchards should be sterilized** using chemical sterilants @2.5-5% formalin or 6 weeks of soil solarization using 50-100µ thick linear low density polyethylene (LLDPE) sheet during hot summer months. If formalin is used ensure that the soil is free from any formalin fumes before transplanting in the bags. Soil solarization is beneficial as it kills harmful pests and pathogens and also increases population of beneficial microrganisms which are present in the soil and are thermo tolerant, whereas, formalin treatment kills both harmful and beneficial organisms.The above formalin treatment can be also be used for sterlizing soil after removing dead plant.

3. On observing first symptoms of wilt first ascertain the cause/s. If it due to **fungal pathogens** in the orchard immediately drench soil with propiconazole 25EC (2ml/l) + chlorpyriphos 20EC (2.ml/l) or carbendazim 50WP (2.0g/l) + chlorpyriphos 20EC (2.ml/l) use 5-10 l solution/plant depending on growth so that 12 inches depth below shaded area becomes wet. Also drench at least

3-4 healthy plants on all the four sides around the infected plant/s, repeat the drenching 3-4 times at 20-25 days interval. Drenching with Ridomil, metalaxyl or dithane M-45 (2g/l) will be beneficial if *Phytophthora* is causing any loss.

4.   For controlling **shot hole borer** (*Xyleborus* spp.) which is associated with wilt disease, 10 litres preparation containing red soil (4kg) + Chlorpyriphos 20EC (20ml) + Copper oxychloride (25 g) needs to be applied on plant base up to 2 ft. from second year onwards. To control stem borer, inject in the holes on the trunk with DDVP 2-3 ml and plug the holes with mud.

5.   Wilt due to **root knot nematodes** can be managed with soil application of phorate 10G @10- 20g/plant or carbofuran 3G @ 20-40g/plant or Fipronil0.3G 30-40g/plant, other suitable nematicidein the plant basin, in a ring near root zone and cover it with soil. Drenching with azadirachtin (1%) @ 2ml/l is also recommended. Application of neem cake 1-3kg/plant depending on age is advisable twice a year. Plant *Tageteserecta* (African marigold varieties best followed by French marigold) between plant to plant space in a row, or in a ring, on the border of plant basin. For effective results these should be grown for more than 4-5 month. Crops like onion, tomato, chili, potato, capsicum, gram, legumes, cucurbits, Gerbera, Gladiolus *etc.* aggravate nematode infestation and hence should be avoided as intercrop. Green manuring with *Sesbania*and sunhempis beneficial.

Biological Control: Biological formulations if used should be reliable, fresh and used during rest period when no other fungicides/bactericides are used. The soil application of *Bacillus subtilis, Paecilomyceslilacinus, Pseudomonas fluorescens, Trichoderma harzianum, Aspergillus niger* AN 27@ 10-15g/plant along with well-decomposed farm yard manure around the trunk of pomegranate trees helps to prevent wilt infections. Neem cake @ 2-3 kg/ plant effectively checks incidence of wilt complex.

Biofertilizer – Kalisena SA having *Aspergillus niger* AN27@1 kg/acre+ Mychorrhizal preparation AMF@ 5kg/acre gives effective control of wilt if use from beginning or before disease starts. These two biofertilizers should be applied twice a year along with sufficient organics for effective wilt management. These controls several soil pathogens and also improves nutrient uptake and gives disease resistance and improves yields.

General Precautions

1.   Once disease is detected in the orchard, dig about 3-4 feet long trench between the wilted and healthy plant/s. The partially wilt affected plant/s should be treated with a suitable soil application depending on pathogen involved.

2. Dead plants should be removed and burnt, they should not be kept dumped in the orchard for firewood. While removing the wilted plants from the orchard for burning, protect the entire root zone with cover- fertilizer bag, *etc.*, so that pathogens insoil on root do not spread in orchard.

3. The soil in the pit from which dead plant has been removed, should be sterilized with 2.5-5% formalin using about 10 l solution. It should be covered with polyethylene sheet for 1 week. After 1 week remove polyethylene sheet and rake the soil daily up to 10-15 days, so as to allow escape of gas. Plant new sapling once there is no smell of formalin in soil.

4. Pruning tools should be disinfected and cut ends painted with fungicidal oil based paints. Pruning should be avoided during spring to summer and done in winter months. Partially affected plants within the buffer zone should be treated with a suitable treatment; neighboring asymptomatic apparently healthy plants should also be treated with appropriate systemic fungicide/ insecticide. Plants with more than 30% canopy loss should not be treated, they should be uprooted and burnt, soil treated with formalin and new plant grown.

30

Status of Major Borer Pests of Pomegranate and their Integrated Management

[2]Sachin S. Suroshe, Mallikarjun [1]M.H and [1]R.K. Pal

[1]ICAR-National Research Center on Pomegranate, Solapur-413 255, Maharashtra
[2]Biological Control Laboratory, Division of Entomology, ICAR-IARI,
Pusa, New Delhi-110 012

Introduction

Pomegranate farming in India is passing through major challenges like climate change, Non availability of resistant cultivars, increased incidence of insect pests and diseases and availability of very scanty information on pomegranate. Recently, the cultivation of high yielding pomegranate varieties with intensive care and management under irrigated conditions has led to emergence of certain severe pest problems in Pomegranate. Pomegranate crop in India is reported to be infested by more than 91 insects, six mites and one snail pest. Pomegranate butterfly, *Deudorix isocrates* Fabricius is considered as the most damaging pest which causes more than 50% losses in the yield. Next in the order of severity are the three species of fruit sucking moths *Eudocima materna* (Linn.), *E. fullonia* (Clerk) and *E. homaena* (Hub.). The bark eating caterpillar *Indarbela* spp. and shot hole borer *Xyleborus fornicatus* and *Xyleborus perforance* are also assuming a status of major pests recently in most of the pomegranate growing states of India. The major pests damaging pomegranate are listed in the table 30.1.

Table 30.1: Major pests attacking pomegranate crop in India

Common name	Scientific name	Family	Order
Borer pest complex			
Fruit borer/ Anar butterfly	Deudorix isocrates (Fabricius)	Lycaenidae	Lepidoptera
Stem borer	Coelosterna spinator Fletcher	Cerambycidae	Coleoptera
Shot hole borer	Xyleborus fornicates (Eichh.)	Scolitidae	Coleoptera
Bark-eating Caterpillar	Inderbela quadrinotata (Walker)	Metarbelidae	Lepidoptera
Fruit piercing moths	Eudocima materna (Linn.) E. fullonica (Clerk) E. homaena (Hub.).	Noctuidae	Lepidoptera

Fruit Borer

The most obnoxious enemy is pomegranate/*anar* butterfly, which may destroy more than 50% of pomegranate fruits .The farmers suffer loss in terms of quantity and quality of fruits. There are two species of fruit borer damaging pomegranate crop in India; *Deudorix epijarbas* Moore is found in hilly states like Himachal Pradesh and Jammu & Kashmir and *Deudorix isocrates* Fabricius in other plain states of India. Biology, nature of damage, and management remains the same for both the species. The pest attacks the pomegranate crop during rainy, winter and summer season crops. Recently it is also becoming serious pest of guava in Uttar Pradesh and other parts of North India. It has spread widely from India to Sri lanka, and Burma. It is found in plains and also occurs to an altitude of 2000m. In India it is present everywhere except deserts. It is recorded as pests of Guava, Orange, Mandarin, Lemon, Aonla, Mulberry, Pomegranate, etc. It is the most important insect pest distributed all over India. The adult males are glossy bluish and brownish violet and in case of females a conspicuous orange patch on the forewings is seen. The adult female lays shining white eggs singly on calyx, bud, flower and young fruits. Incubation period lasts for 7-10 days and on hatching, the caterpillar bores into fruit and feed on the pulp. The damage of fruit borer is seen throughout the year irrespective of the *bahar*. Larval period lasts for 18-47 days. Larva is dirty brown, strongly built, covered with short hairs and measures up to 20 mm long. Pupation occurs either inside the damaged fruits, on the fruit surface or on the ground for 7-34 days. Total life cycle is completed in 1 to 2 months. Round holes can be seen on bored fruits, from which larval excreta comes out continuously. The injury caused by fruit borer attracts bacteria, fungi and beetles. The activity of the pest remains lowest during March to June, while the period from July to October is most favourable for its build-up and the period from November to February remains intermediate. Application of physical barrier such as covering the (30-40 day old) fruits with Poly Propylene non-woven white bags and butter paper bags provides total protection against insect pests. Removal and destruction of all the damaged fruits showing exit holes. Clipping off calyx cup of flowers immediately after pollination helps to reducing the load of fruit borer eggs on the fruits and thus avoid the further damage. Application of Five spray of fenvalerate 0.1% or fluvalinate 0.01% or carbaryl 0.2% or triazophos 0.05% at interval of three weeks commencing at the initiation of fruit setting works well. Two rounds of spray with neem oil at 3 percent also control borer infestation effectively. Release of *Trichgramma chilonis* @ 2.5 lakhs/ha four times at 10 days interval reduces fruit borer infestation. Recently, a new species of egg parasitoid, *Trichogramma manii* has been reported from eggs of pomegranate fruit borer, *Deudorix isocrates* which can be further exploited for augmentative biological control.

Stem Borer

Stem borer, once considered minor pests of pomegranates assuming major status, and is becoming dreaded in combination with shot-hole borer. Cerambycid beetle, *Coelosterna spinator* is normally found associated with pomegranate. Adults of Cerambicidae are referred as long-horned beetles because of their long antennae. Stem borer, *C. spinator* bores the stems and trunk of pomegranate trees. It is polyphagous pest became the major pests of fruit crops. Adult is about 4 cm long and dull yellow with minute spots. Capsule shaped eggs are laid singly in each of the slits made by the adult females and are covered with a hard gummy substance. Egg period is 12 to 15 days, grub 9 to 10 months and pupal period is16 to 18 days. On hatching, the flat-headed cream coloured grubs bore and make tunnel into the heartwood of the tree. The grating and scraping sounds of grub can be heard quite distinctly if you place your ear against the trunk of an affected tree. The adult emerge from infested plants during July to September through a round hole made by chewing wood. Adult beetles are active by the day and feed by gnawing the green bark of shoots. It normally completes one generation per year and longevity of beetles is 45 to 60 days. It prefers breeding in dead wood but also attacks the living branches. Presence of saw dust like pellets under the tree indicates the damage is being caused by the grub of borer. Leaves of damaged branches turn yellow and subsequently get weakened, which badly affect the growth of trees. The maturity of pomegranate fruits is also gets delayed, affecting the yield and quality. The best way to control stem borers is to prevent plants from becoming weak due to poor nutrition, attack by other insect pests, pH imbalance, and excess/deficiency of water. A healthy tree is more likely to fend off stem borer attack. Hand picking and killing of adults in kerosene water from July to September can be performed. Mechanical killing of eggs and first instar grubs hiding in the slits with the help of a sharp knife from July to September is also effective. The grubs of SB can be stabbed by inserting a hooked wire down the hole. Pruning of dead branches during pre-monsoon and before taking *bahar* helps in keeping adults beetles at bay. Do not pile dead or uprooted wood within or near orchards as SB beetles breed on it and attack nearby pomegranate orchards; burning and destroying is the best practice. Adult beetle can be attracted and get killed by installing the light traps with 200 watts sodium bulb immediately after the first showers/rain Stem injection with dichlorvos 76 EC @ 80 ml/litre results in the 90 to 100% reduction of live holes caused by *C. scrabrator*. Dichlorvos solution has to be applied with the help of squeeze bottle till the tunnel is filled and then closing with the wet mud. Stem borer damage can be prevented by pasting prepared out of 10 litres of water + red soil (4kg) + chlorpyriphos 20EC (20ml) + copper oxychloride (25 g). It needs to be applied from plant base (Collar region), with the help of brush, up to 1-2 feet above from second year crop onwards; it discourage the oviposition and boring by beetles.

Shot Hole Borer

Shot-hole borer (SHB) is becoming most serious pests of pomegranates grown under heavy soils. The damage is caused by constructing the galleries in the branches and subsequently the accumulation of wood rot. The SHB is very difficult to control due to its well concealed habitats. Two species of shot-hole borer, *Xyleborus fornicatus* (Eichh.) and *Xylosandrus compactus* (Eichhoff) are causing problems in pomegranate. The beetles can reproduce on a range of host plants belonging to Leguminosae, Verbinaceae, Moraceae, and Euphorbiaceae. The adult beetles are black in colour and cylindrical in shape. Females bore galleries inside the stem and infest it with the spores of ambrosia fungus, *Monacrosporium ambrosium*. SHB infestation is a combined attack of the beetle and the ambrosia fungus and it leads to death of the pomegranate plant. SHB remains active throughout the year with higher activity during the post-monsoon period. Exit holes or shot holes from which adult of SHB come out can be seen on collar region as a mark of symptoms of attack.

Easy methods of control include cutting, burning and destruction of pruned infested branches. Removal of alternate host crop castor from vicinity of orchards checks the X. fornicatus populations. Do not pile dead or uprooted wood within or near orchards as beetles breed on it and attack nearby pomegranate orchards. SHB damage can be prevented by pasting prepared out of 10 litres of water + red soil (4kg) + chlorpyriphos 20EC (20ml) + copper oxychloride (25 g). Prepared paste is applied from plant base (Collar region) up to 1-2 feet above from second year of crop onwards for discouraging the oviposition and boring by adults.

Bark Eating Caterpillar

It is a pest of minor importance, but quite frequently encountered, and always found to be associated with neglected, unmanaged and new orchards. Bark eating caterpillars are polyphagous pest and found associated with trees and shrubs. This is distributed from India, mainland South East Asia and China. Several species of *Inderbela* are reported as pests of many trees in both Africa and Asia. Two species, *Inderbela quadrinotata* (Walker) and *Inderbela tetraonis* (Moore) are serious pests of mango, ber, citrus, fig, guava, sapota, phalsa, Indian blackberry, litchi, mulberry, pomegranate, drumstick, loquat, aonla, sea-oak, copper-pod, umbrella tree, curry leaf, amaltas, almond, portia tree, goldokan, saru, caoba, kenda, cashew nut and kapok throughout India. Members of Metarbelidae are termed as Wood borer moths, the adults and larvae are nocturnal. Male length varies from 22-25 mm, while females are 26-29 mm. The male adults are whitish grey, stout with brownish streaks on pale whitish forewings and hind wings are pale. The female moths are pale brownish with forewings having row of dark rusty red spots. Eggs are laid in clusters of 15-20 directly on the bark of branch. The freshly laid eggs are oval in shape, pale colored which later turn creamy white. Eggs hatch in 8-10 days after being laid and emerged larvae are 1.6cm; whereas, fully grown larvae are up to 5cm long. Pupation takes place in the tunnel in the wood by larvae. The pupa possesses rows of teeth or hooks on the abdominal segments by means of which it climbs out of the tunnel. The pupal period lasts for about 15-25 days. Larva bores a tunnel downwards in to the wood, usually at the junction of branches and feeds on the bark of the tree at night for 9-11 months. Wood dust and faecal matter hangs in the form of a web around the affected portion. The tunnels causes weak points on the tree, where breakage occurs, affecting the vitality of the trees badly. Life cycle is annual with one generation per year. The initial infestation of this insect occurs when the trees are relatively young, so regular inspection of orchard is must to detect the damage at the beginning. Poking a hooked wire into the bored hole kills the larvae inside, it can be quite effective in small orchards or when the infestation is low. Keeping the orchards clean and avoiding overcrowding of trees help to minimize the attack. Clean the web and excreta of the larva hanging near the affected portion, as the larva hides beneath it and feeds on the bark. This pest prefers the badly pruned trees, so proper pruning i.e. without mechanical injury to trees is the best preventive measure. The pruned dead branches should be destroyed by burning. Injecting carbon disulphide or paradichlorobenzene in the tunnel, and sealing it with the mud kills the larvae. Injecting fenvalerate (0.01%) and quinalphos (0.05%) into larval holes is effective in controlling the bark eating caterpillar. Spraying of insecticides like dichlorvos (0.08%) and monocrotophos (0.08%) also gives good control of bark eating caterpillar. Do not pour petrol or kerosene into the holes as these products are very damaging to the bark and wood of the tree and the tree will take at least three times longer to recover.

Fruit Piercing Moths

Fruit orchards all over the world are attacked by the caterpillars of Lepidoptera, but in case of fruit piercing moths (FPM) damaging stage is an adult. All species of moths recorded piercing the fruits are belonging to the family Noctuidae. The fruit-piercing moth attacks many fruit and vegetable crops *viz.*, apple, apricot, banana, breadfruit, coffee, fig, grapefruit, guava, kiwifruit, litchi, longan, mandarin, mango, nectarines, orange, papaya, passion fruit, peach, persimmon, pineapple, plum, pomegranate, star fruit, tomato and melon. Three species of moths i.e. *Eudocima materna* (Linnaeus) *E. fullonia* (Clerck) and *E. homaena* (Hubner) are found infesting pomegranate fruits of *mrig bahar* in different parts of India. *E. fullonia* is covered with orange colored scales and an inverted "C" shaped marking is seen at the centre of hind wings in both the sexes. *E. materna* is covered with orange colored scales and a black dot marking is seen at the centre of hind wings in both the sexes. *E. homaena* is covered with orange colored scales and fore wings are with parrot green color band, and an inverted "C" shape marking is seen at the centre of hind wings in both the sexes. FPM remains active from August to October and only attacks the fruits of *mrig bahar* crop of pomegranate. FPM generally remain absent in the locality of fruit crop for years, and then suddenly destroy orchards in few days during nights. Moths of genus *Eudocima* have a highly specialised proboscis with hard spines capable of piercing hard and unripe fruits. Adult males and females of moth penetrates the fruits at night and damaged fruits become soft owing to secondary infections from different fungi and bacteria. Punctured holes on fruits with oozing fruit juice can be noticed.

FPM/FSM attack occurs along the outer rows of the orchard. So, fruit crops should be planted in square blocks and not in a few long rows. Collection and destruction of moths using torch in the night is the best and efficient practice. As the activity of fruit sucking moth remains from August to October avoid taking *mrig bahar* in pomegranate. Cover the fruits with butter paper or Polypropylene Non-woven bags as per the availability of covering materials. Destroy hosts plants of larvae *viz.*, *Tinospora cordifolia, T. sinensis, T. smilacina, Cocculus hirsutus, Anamirta cocculus, Tiliacora acuminate, Diploclisia glaucescens, Cissampelos pareira, Convolvulus arvensis, Trichisia pattens and Pericampylus glaucus* present in and around the field. Smoke masks the odour of the mature fruit, so that, the moths fail to detect fruits and do not enter the orchard. Egg parasitoids, *Telenomus* sp., *Ooencyrtus* sp. and *Trichogramma* sp. are among the most successful biological control agents of FPM. Poison bait (95% molasses/jaggery + 5% malathion), attracts adult FSM and killed them through stomach action.

31

Major Sucking Pests of Pomegranate and their Integrated Management

[1]Mallikarjun M.H., [2]Sachin S. Suroshe and [1]R.K. Pal

[1]ICAR-National Research Center on Pomegranate, Solapur-413 255, Maharashtra
[2]Biological Control Laboratory, Division of Entomology, ICAR-IARI, Pusa,
New Delhi-110 012

Introduction

Cultivation of high yielding varieties of pomegranate with intensive care and management in the recent past under irrigated condition with early stage exploitation of plant has led to certain severe pest problems. Among them, infestation by sucking pests like whitefly, *Siphoninus phillyreae* (Haliday); mealybugs, *Pseudococcus lilacinus* (Cockerell) and *Phenacoccus solenopsis* Tinsley; thrips, *Scirtothrips dorsalis* and *Rhipiphorothrips cruentatus* Hood; aphid, *Aphis punicae* (Passerini) scale insects attack the crop during new flesh, flowering and fruiting stages and results in reduction of pomegranate fruit yield and put the growers into hardship. The growers loose in terms of quantity and quality of fruits also. The major constraint in increasing export potential is the quality of fruit in terms of size, colour, free from blemishes and pesticide residue blow MRL levels. To overcome the latter constraint it is necessary to follows the eco-friendly management practices for sucking pests.

Common name	Scientific name	Family	Order
Sucking pests			
Ash Whitefly	Siphoninus phillyreae (Haliday)	Aleurodidae	Hemiptera
Aphids	*Aphis punicae* (Passerini)	Aphididae	Hemiptera
Mealybugs	*Pseudococcus lilacinus* (Cockerell) *Planococcus citri* (Risso) *Planococcus lilacinus* (Ckll.) *Ferrissia virgata* (Ckll.) *Phenacoccus solenopsis* Tinsley	Pseudococcidae	Hemiptera
Thrips	*Scirtothrips dorsalis* Hood *Rhipiphorothrips cruentatus* Hood	Thripidae	Thysanoptea

Ash Whitefly

The adult ash whitefly (*Siphoninus phillyreae*) is a pale whitish with slightly mottled wings. The winged adults fly makes irregular motion like tiny moths when disturbed. Ash whitefly is native to Europe, the Mediterranean and Northern Africa, and now it is found in many other countries. Females may live up to thirty to sixty days, while males live an average of 9 days. A winged female lays eggs on the underside of the leaves. When the nymphs emerge they do not move far and feed on the plant sap by digging their mouth parts into the leaf tissue for sucking the sap until pupation. On first observation, the pupal case appears similar to the white encrustation of snow scale. The pupal case is 0.8-1.0 mm long and 0.55-0.7mm wide. The entire life cycle from egg to adult usually takes place under the same leaf. Serious damage is caused by the excretion of honeydew by whitefly. Under moist conditions sooty mould develops on honeydew reducing photosynthesis and respiration of plants. Curling of leaves and growth of black sooty mould can be seen on the tender leaves. Heavy infestations cause leaf yellowing, early leaf drop and smaller fruits. It can be managed by spraying water with high volume sprayer helps in washing out the whiteflies followed by spray of triazophos 40 EC @ 1.5 ml/l.

Aphids

Aphids (*Aphis punicae*) are yellowish green in colour with two cornicles at the dorsum of last abdominal segment. Generally aphids can be seen in pomegranate orchards during winter *i.e* from November to March and mostly damages *hasta bahar* crop. It is distributed throughout the Mediterranean region, Middle East, Ethiopia, India and Pakistan. The optimal temperature for aphid growth, development, and reproduction is between 22-25 °C. High humidity favours the multiplication of aphids. The average longevity of adult females is about 8 days. The average numbers of offspring produced by a single female are 11.27 at temperatures of 27.5⁰C. Nymphs and adults suck the sap from tender shoots, leaves, flowers, buds and fruits. It excretes sweet semisolid sugary substance called 'honeydew' which attracts black sooty mould growth. It is a serious problem on new flush. The affected leaves show chlorotic patches. Spraying with dimethoate 30 EC @ 2 ml/litre or acetamiprid 20 SP @ 0.5 ml/l or imidacloprid 17.8 SL @ 0.5 ml/l at 15 days interval effectively manage the aphid population. Release of predators namely *Scymnus sp, Menochilus sexmaculata* helps in bringing down the population of *A. punicae* on pomegranate

Mealybugs

Six species of mealybugs are recorded from pomegranates in India Oriental mealybug, *Pseudococcus lilacinus* (Ckll) Citrus mealybug, *Planococcus citri* (Risso) Coffee mealybug, *Planococcus lilacinus* (Ckll.) Stripped mealybug, *Ferrisia* virgata (Ckll.) Cotton mealybug, *Phenacoccus solenopsis* Tinsley Vine mealybug, *Planococcus ficus* (Signoret). Females are oval with waxy coatings all over the body and males are winged. Mealybugs pass through 4 female and 5 male instars. Mealybugs, depicts diverse range of life history parameters thus any generalised life history cannot be arrived at. Eggs or crawlers (1st instar) are laid by the adult female. Eggs are usually laid in a bag called 'ovisac', species that lay crawlers rather than eggs lack ovisac. Even though many species have legs, most mealybugs remain stationary throughout their life. Most species have 1 or 2 generations a year, although some are reported to have as many as 8 generations in the greenhouse. Both parthenogenetic and sexual reproduction is seen in mealybugs. Leaves show characteristic curling symptoms much like a viral infestation. Black sooty mould develops on excreted honeydew. The infestation may lead to fruit drop. The market value of heavily infested fruits is reduced considerably. The plants in the vicinity of the orchard acting as alternate hosts for the mealybugs should be destroyed. Keep the orchards weed free as the Parthenium hysterophorus serves as an alternate host for Phenacoccus solenopsis. Pruning of infested twigs, fruits will help in minimising the spread and infestation of mealybugs. Chemical: Spray chlorpyriphos 20 EC @ 2.0 ml/l or dimethoate @ 2.0 ml/l or malathion 50 EC @ 2.0 ml/l with fish oil rosin soap. Mealybugs are naturally controlled by ladybird beetle predators. The species of coccinellid that exercise natural control over mealybugs are *Brumoides suturalis* (Fabricius), *Cheilomenes sexmaculata* (Fabricius), *Coccinella septempunctata* Linnaeus, *Nephus regularis Sicard* and *Scymnus coccivora*. An encyrtid parasitoid, *Aenasius bambawalei* has been reported to control almost 70% of natural infestations of Phenacoccus solenopsis under field conditions

Thrips

S. dorsalis is commonly known as chilli thrips in India. Nearly 112 taxa of plants are recorded as a host of this species. It is a pest of various vegetable, ornamental, and fruit crops in southern and eastern Asia, Africa and Oceania Thrips are the most serious sucking pest attacking pomegranate during flowering and fruiting stages. Generally hot and dry climate or prolonged dry spells after start of monsoon shower favours the multiplication of thrips. New shoot and leaf growth after pruning are more susceptible to thrips attack. The adults of *S. dorsalis* are straw yellow coloured whereas the nymphs are minute yellowish brown in colour. Reports from India indicate damages by the thrips, *S. dorsalis* on pomegranate. The adults of *R. cruentatus* are minute sized (measure 1.4 mm long), slender, soft bodied insects with heavily fringed blackish brown or yellowish wings. Females lay about 50-70 bean shaped or elongate eggs on the under surface of tender leaves and floral buds. The incubation period is about 3-5 days. Newly hatched nymphs are reddish and turn yellowish brown as they grow. There are total four nymphal instars with first two stages as active feeders and the last two as non-feeding, quiescent prepupal and pupal instars. The prepupae and pupae can be seen on under surface of the leaves. Pupal period lasts for about 2-5 days. Average life cycle from egg to adult is completed in about 14-15 days. Thrips can complete as many as eight generations in a year. Thrips are characterised by possessing rasping and sucking type of mouth parts with right mandible absent. Both the nymphs and adults damage the crop by lacerating the leaves, flowers and developing tender fruits, and suck the oozing sap. This results in curling of leaves, heavy shedding of flowers and buds, drying of growing tips and appearance of rusty scars or scabs on the fruits. The thrips damage on fruits badly affecting their market quality. Remove and destroy affected plant parts and tender new growth promptly from time to time. Do not intercrop other host crop of *S. dorsalis* such as chilli, onion and garlic in and around or near the pomegranate orchard. Managing the attack of thrips in high value fruit crops like pomegranate is crucial though quite cumbersome task. In order to comply with international standards for pesticide residue levels and to minimise the associated health hazards, judicious use of selective chemicals is imperative in production of the export quality pomegranates. Spraying thiamethoxam 25 WG @ 0.3 g/l or acetamiprid 20 SP @ 0.3 g/l or acephate 75 SP @ 1 g/l at flowering and fruit setting stage gives optimum control. Repeated use of same pesticide for the long duration might leads to development of resistance in thrips, so it has to be alternate with other efficient pesticides.

32

Non-Insect Pests of Pomegranate and their Integrated Management

[1]Mallikarjun M.H., [2]Sachin S. Suroshe and [1]R.K. Pal
[1]ICAR-National Research Center on Pomegranate, Solapur-413 255, Maharashtra
[2]Biological Control Laboratory, Division of Entomology, ICAR-IARI, Pusa,
New Delhi-110 012

Introduction

Intensive cultivation of high yielding varieties of pomegranate with improved care and management in the recent past under irrigated condition with early stage exploitation of plant has led to certain severe pest problems of certain mite and nematode pests like *Tenuipalpus punicae and Aceria punicae* Pritchard & Baker, *Melodogyne incognata* Wartellei. Mite pests attack the crop during new flesh, flowering and fruiting stages and results in poor fruit set and reduction in marketable quality of fruits. The major constraint in increasing export potential is the quality of fruit in terms of size and colour. Nematodes attack on the young roots of the plants of all stage and leads to the stunted growth and wilting of the plants in progressive stage. The major pests damaging pomegranate are listed in the table 32.1.

Common name	Scientific name	Family	Order
Nematode pest			
Root knot nematode	*Melodogyne incognata* Wartellei	Heteroderidae	Tylenchida
Mite pest			
False spider Mite	*Tenuipalpus punicae, Tenuipalpus granati* Pritchard & Baker	Tenuipalpidae	Acarina
Eriopid mite	*Aceria punicae,*	Eriophyidae	Acarina
Tetranychidae Mite	*Aceria granati*	Tetranychidae :	Acarina

Nematode

Root-knot nematodes belong to the genus *Meloidogyne,* and are plant-parasitic in nature. About 2000 plants are susceptible to infection by root-knot nematodes and they cause approximately 5% of global crop loss. The genus includes more than 60 species, with some species having several races. Five Meloidogyne species (*M. incognita, M. javanica, M. arenaria, M. hapla* and *M. acrita*) have been reported from pomegranate. *M. incognita* has very broad host range *i.e.* more than 700 hosts. The situation regarding *M. incognita* is complicated by the existence of morphologically indistinguishable races differentiated by their ability to reproduce on a range of test host plants *M. incognita* is widely present in pomegranate growing regions of the country, and race 2 of this species is causing substantial yield losses and also reducing the quality of pomegranates. The life cycle of root-knot nematode is largely indifferent with respect to individual species host-parasite relationships and physiological characteristics. Adult female lays hundreds of eggs which pass through an embryonic stage, four juvenile stages and an adult stage. Adult females (0.4-1.3 mm long) are pear-shaped and always lay eggs in gelatinous matrix. Incubation period for eggs is about 7 days. Second instar vermiform juveniles (0.3-9.5 mm long) hatch from the eggs and invade roots at the elongation region (root cap). This second stage infective juveniles migrate between and through cells, and position themselves with the head in the vascular tissues. Due to internal migration by juveniles cell gets damaged and subsequently cell division stops. As feeding continues several cells enlarge and become multinucleate and are called as giant cells or galls. These changes are induced by salivary secretions introduced into cells and tissues by juveniles during the feeding. As the xylem vessel gets damaged the upward movement of water and nutrients halt and disrupt the growth of plant. Infested plants reveal pale green or yellowish leaves with reduced plant growth and eventually results in wilting and death of the plants. The entire life cycle is completed in 22-28 days. Males are vermiform, and are not required in reproduction (parthenogenic). Root-knot nematodes can be spread by water or by soil which clings to the farm equipment or through infested planting materials. Deep summer ploughing of nursery and main field before planting of grafted seedlings and stem cuttings is required for reducing the inoculum. Soil solarisation of main field and nursery reduces the nematode populations in the top 35 cm of soil. First moisten the soil with water and then cover it with clear transparent polythene sheet. Cover the edges of polythene with layer of soil, so that it becomes air tight. Leave it for 30 to 45 days during the hottest part of summer in May. Root knot nematodes eggs and infective juveniles die when the soil temperature covered with polythene exceeds 52⁰C for 30 minutes or 55⁰C for 5 minutes. Keeping the land fallow for at least for six months to one year followed by one or two deep summer ploughing of nursery and main field before planting is required for reducing the inoculum. First moisten the soil with water and then cover it with clear transparent Polyethylene sheet and Leave it for 30 to 45 days during the hottest part of summer in May. Soil solarisation of main field and nursery reduces the nematode populations in the top 35 cm of soil. Use the certified nematode free seedlings or tissue cultured plants for planting. Inspect the presence of root galls by uprooting the few pomegranate seedlings from polythene bag in the nursery and if you found the galls on root do not use such seedling for planting and immediately destroy by burning. Application of FYM/Compost/Chicken litter at 20-25 t/ha or non-edible oil cakes (Pongemia or Neem oil cake) at 2 t/ha in the pomegranate nursery as well as in the main field at planting and during *bahar* treatment every year serves well in reducing the root knot nematode population. Intercropping pomegranate with sun hemp, marigold (*Tagetes erecta, T. patula*), mustard, during rest period also restricts the nematodes. Maintaining the nursery area or main filed free from the weeds as the weeds serves as the alternate host for nematode survival and perpetuation. Removal and destruction of the infested planting materials at nursery and infested plants in main filed need to be destroyed by burning. Soil application of carbofuran 3G or phorate 10G at 2 kg a.i. /ha in the nursery as well as during planting in the main field and at 4 kg a.i. /ha during bahar treatment every year checks the nematode population. Application of Neem cake @ 5-10kg per plant repels the nematodes from the plant vicinity. Application of bioagents *viz., Trichoderma viride, Paecilomyces lilacinus, and Pseudomonas fluorescens @ 10 to 20g /m2* nursery or @ 5 kg with 100 kg FYM/Compost/ha in the main field every year during planting and *bahar* is effective against root nematodes. Hence, there is a need to develop Integrated Nematode Management (INM) practices for the effective management of nematodes.

Mite

Mites are nothing but tiny spiders. The members of *Tenuipalpidae* are often called as false spider mites because they do not construct the web and are reddish in color like red spider mites. *T. punicae* was first described by Pritchard and Baker (1958) and reported on pomegranate first time at Ludhiana in 1972. These are very small (0.2-0.3 mm long) and reddish in colour. A suture separating the propodosoma from the hysterosoma may or may not be present. The adults usually have four pairs of legs. Males are very rare and females reproduced parthenogenetically. Temperature plays an important role in the development of this mite. It is most active during dry spell. Adult and nymphs feed on the lower leaf surface by scrapping leaf and sucking the liquid contents. They can be easily identified due its red colour. If you press your thumb against the surface of infested leaves your thumb turns red. Shiny white/silvery patches can be seen on the under surface of affected leaves which may further curl and fall. Do not neglect the orchards during the off period and must provide irrigation frequently to avoid dryness. Spray dicofol 50 WSP @ 1 g/l or sulphur 80 WP @ 2 g/l/ Abamectin 1.9 EC 1ml/l, Propergite 50% EC 1.5 ml/l/ Chlorfenpyr 10% SC 1.5g/l/ Diafenthurion 50% WP 0.75ml/l/ Milbemectin 1% EC0.3ml/l/ Fenazaquin 10 % EC 1.5ml/l/ Fenproximate 5% EC 0.4ml/l/ Phasalone 35 % EC 2ml/l/ NSKE 5 %. Repeat the spray after fortnight if mite population does not decline with any of the alternative Miticides/insecticide.

33

Post Harvest Management of Pomegranate Fruits

D.V. Sudhakar Rao and C.K. Narayana

Division of Post Harvest Technology
Indian Institute of Horticultural Research, Bengaluru-560 089, Karnataka
E-mail : cknarayana2001@yahoo.com

Pomegranate (*Punica granatum* L.) is one of the hardiest fruit crops that grow well under arid and semi arid climatic conditions where winters are cool and summers are hot and dry. The arils of fruit are sweet acidic in taste and are used both for dessert purposes as well as for its refreshing juice. The varieties of pomegranate vary in their size (6.25mm to 12.5 mm in dia), weight (120 g to 400 g), taste (sour to sweet) and colour (pinkish white to deep red). Recently pomegranate is gaining popularity due to its health benefits.

Several pre-factors contribute to postharvest losses in pomegranate unlike other fruits and vegetables. However, the benefits of well managed production system may be negated due to poor postharvest handling leading to loss of this precious commodity. During fresh fruit marketing, pomegranate fruits become dull, desiccated, tough, deformed and discoloured beyond one week after harvesting, reducing its marketability and price. The post harvest losses of pomegranate could be reduced significantly by adopting proper post harvest management practices during picking, handling, packing, storage, transportation and marketing.

Maturity

Being a non-climacteric fruit pomegranate has to ripen on tree and must be picked at right maturity stage to ensure the best eating quality. The fruits when fully ripe have a waxy shining surface of red or yellow peel colour, depending on the cultivar. The fruits must not be left to over mature, as it tends to crack open if rained upon or under certain conditions of atmospheric humidity, dehydration by winds, or insufficient irrigation. At the same time early harvesting will lead to poor quality. Generally the fruits will be ready for harvesting after 5-7 months of flowering. The harvest maturity is characterized by red or yellow colour of peel (depending on cultivar) without any greenish tinge, red colour of arils/juice and a TSS content of juice around 15 to 16 %.

Harvesting

The peel being high in phenolics, fruits should be harvested carefully by either hand picking or using the clippers or secateurs. Care should be taken to avoid any physical injury to the fruits during harvesting and should be done in the cool hours to avoid the effect of high temperature. The harvested fruits should be placed carefully in the clean containers preferably plastic in crates to avoid mechanical injuries and for easy handling. Sunburn can cause fruit injury leading to a leathery, tough area on rind, usually at the stem end. The injury may show only on one side or may extend completely around the fruit.

Sorting & Grading

Sorting of fruits in pomegranate is mainly done to remove rotten, diseased, mechanically damaged, cracked and insect infested fruits. Proper grading and packing fruits are important unit operations to maintain the quality and to fetch optimum price in domestic as well as international market. In India there is no specific grading parameter for pomegranate, however, the grading is normally done on the basis of size, shape, colour, weight and appearance of the fruit. The fruits are graded as Extra large, large, medium and small according to their size. On the basis of fruit weight they are graded into the following three grades.

Sl. No.	Grade	Fruit weight (g)
1.	Grade A	350 and above
2.	Grade B	250 to 350
3.	Grade C	< 200

Packing

Pomegranate fruits are now being sent to the market in CFB boxes as the growers are aware about importance of packaging. The size of box depends on the size of the fruits packed. Twelve fruits of 300 - 350 gram weight can be accommodated in 30 x 40 x 10 cm size CFB box. In Maharashtra state the growers pack the pomegranate fruits in CFB boxes of following dimensions:

Size of fruit	Size of CFB box	No. of fruits
Extra large	12 x 8 x 4.5"	6
Large	15 x 11 x 4"	12
Medium	14 x 10 x 4"	12
	13 x 9 x 3.5"	12
Small	No specific size	24 or above

In Karnataka, the fruits are generally packed in 9 kg capacity CFB boxes in 3 layers of 9 fruits each with paper shreds as cushioning material for local market. The other grades that are in vogue in Karnataka markets are as follows:

Grade	No. of fruits	Gross wt
12 Super	12	6 kg
12 A	12	5 kg
27 A	27	9 kg
27 B	27	7 kg
24 A	24	6 kg
48 F	48	9 kg
Loose	No specific no.	9 kg

For export market the fruits should be wrapped with tissue paper and packed in two rows in CFB boxes of 40 x 24 x 20 cm size or in a single row of 40 x 24x10 cm size boxes. News paper shreds, bubble sheets or polyester foam sleeves can be used as cushioning material. Package inserts like moulded pulp or plastic trays and cells to isolate individual fruit can also be used to avoid any kind of mechanical damage.

Pomegranate fruits become dull, desiccated, tough, deformed and discoloured mainly due to loss of moisture. Flexible film packaging helps in reducing these losses and extends the storage life. Modified atmosphere packaging (MAP) has been found to extend the storage life of pomegranate significantly. In this method pomegranate fruits are placed loosely inside a selectively permeable plastic film and sealed leaving more space surrounding the fruits. Due to natural process of respiration, oxygen is reduced and carbon-di-oxide increased creating modified atmosphere inside the package. In addition humidity also increases inside the packs due to water vapour released by fruit during transpiration process and it also helps in reducing the weight loss.

Storage and Shelf Life Extension

Low Temperature Storage

Low temperature storage between 5-8°C with 90-95% RH is recommended for long term storage of pomegranate. Prolonged storage at or below 5°C was reported to induce chilling injury in several cultivars. The chilling injury symptoms include brown discolouration of the skin, pale colour of the arils, brown discouloration of the white segments separating the arils and increased susceptibility to decay. Scald is another physiological disorder limiting the long term storage of pomegranate fruits. The symptoms of this disorder is limited only to surface and seen more in late harvested fruits.

Modified Atmosphere Packaging Storage

In studies conducted at IIHR, Bangalore, the storage life of pomegranate could be extended up to 3 months by MA packing the fruits in selectively permeable films and storing at 8°C. This method is however not suitable for ambient temperature, due to condensation of moisture droplets inside the packs leading to rotting of fruits. To overcome this problem, a new technology known as Individual shrink wrapping (ISW) has been standardized at Indian Institute of Horticultural Research (IIHR) for extending the storage life of pomegranate fruits for one month at ambient temperature and over three months at low temperature (7-8°C).

Harvest Freshness Maintained in Shrink Wrapped Pomegranate Fruits after 1 Month at RT and 3 Months at 8°C

Controlled Atmosphere Storage (Cas)

Superior advantages have been reported of CAS over normal cold storage in terms of complete arrest or delaying of spread of postharvest diseases and decreased incidence of storage disorders like chilling injury. The composition of O_2 and CO_2

gases in CAS varies depending on the cultivar between 2- 3% and 6-15%, respectively. An extended storage life of 4-6 months has been reported at a temperature of 6-7°C under controlled atmospheric conditions.

Other Pre-Treatments

Other pre-storage treatments like application of polyamines (putricine and spermidine), waxing using different commercial wax formulations, hot water treatment, hot air treatment, intermittent warming, salicylic acid application and 1-MCP have been tried with varying degrees of success either in delaying or preventing the physiological disorders when stored at chilling temperatures.

Export

Considering high keeping quality and availability throughout the year, India has great export potential for pomegranate. Varieties like Ganesh, Mridula, Arakta, Ruby and Bhagwa are quite superior in quality and suitable for export market. UAE, Saudi Arabia, Bangladesh, Bahran, Qutar and Kuwait are major importing countries of Indian pomegranate.

Quality Indices for Export

* Freedom from growth cracks, cuts, bruises and decay.

* Attractive skin colour and smoothness.

* Good Flavour (depends on sugar/acid ratio).

* TSS above 17 % is desirable.

* Tannin content below 0.25 % is desirable.

34

Post Harvest Management of Pomegranate (*Punica granatum* L.)

R.K. Pal and K. Dhinesh Babu

ICAR-National Research Centre on Pomegranate, Solapur-413 255, Maharashtra
E-mail: rkrishnapal@gmail.com

Introduction

Pomegranate (*Punica granatum* L.) is an important fruit crop of arid and semiarid regions of the world. It is a highly remunerative crop for replacing subsistence farming and alleviating poverty. The cultivation of pomegranate by mankind as a fruit crop dates back to antiquity. The usage of pomegranate is deeply embedded in human history with references in many ancient cultures for its use in food and medicine. It is one of the oldest known edible fruits and is associated with ancient civilizations of the Middle East.

The pomegranate is believed to be originated from Iran (Primary centre of Origin). Besides, it is widely prevalent in Afghanistan, Pakistan and India, the Secondary Centres of Origin. From its origin in the area now occupied by Iran and Afghanistan, the pomegranate cultivation had spread to India, China and Mediterranean countries viz., Turkey, Egypt, Tunisia, Morocco, and Spain. Spanish missionaries brought the pomegranate to the America in the 1500s.

Pomegranate fruit has a good consumer preference for its attractive, juicy, sweet acidic and refreshing arils. The ancient fruit has emerged as a commercially important fruit in the recent times. There is a growing demand for good quality fruits both for fresh use and processing into juice, syrup, squash, wine besides anardana, an acidulant. With respect to nutritional content, pomegranate is ranked better than grapes, mango, orange and papaya. Pomegranate is one of the richest sources of Riboflavin. Rind of the fruit, bark of stem and root of pomegranate contain more than 28% gallotannic acid and yellow dye that could be extensively used in tanning as natural bio-dye.

Pomegranate is classified as difficult to eat fruit; hence minimal processing through separation of edible parts (arils) for making it a convenient food is of great importance all over the world. Further, at any given time approximately 60 per cent of harvested fruits fetch the premium price for export, approximately 20-25 per cent of the produce can be sold in the domestic market whereas 15-20 per cent of the harvested produce may not fetch good price due to fruit spots, poor size, shape, colour and other defects. Most of the times taking these fruits to the market does not even meet the expenditure for its transportation. Dumping of these fruits in the compost pit also does not yield good quality FYM due to presence of high polyphenolic compounds that hinders the decomposition process. The ICAR-NRCP has been engaged in various R&D activities on total utilization of pomegranate for production of several value added products *viz.* juice, RTS, wine, seed oil and utilization of rind powder for industrial, pharmaceutical and cosmetic use. Several potentialities of post harvest management and value addition of pomegranate through its total utilization are briefly discussed in this chapter.

Status of Pomegranate Cultivation

Pomegranate grows under a variety of climatic conditions, viz., tropical and subtropical, temperate regions and in hilly areas upto 1800m altitude. It is a deciduous shrub in the temperate region but under tropical and subtropical conditions, it is evergreen or partially deciduous depending upon the variety / genotypes. It grows very well and produces better crop in the semi arid climate where cold winter and hot summer prevails. The plant requires warm and dry climate during the period of fruit development and ripening. Although the pomegranate tree can withstand frost, temperatures below -12°C injure the crop. Further, the crop withstands heat, drought and moisture deficit. The threshold limit for pomegranate cultivation ranges from 44°C to -12°C. It can be grown successfully in arid regions where life saving irrigation facilities are available. The heavy thick rind of the fruit enables it to withstand long distance transport with minimal losses and the fruits can be kept for about 2-3 weeks under ambient condition. Some growers have demonstrated profit upto15 lakh/ha/ annum. It is therefore, highly remunerative crop to replace subsistence farming.

In India, Maharashtra state is considered as the 'pomegranate basket of India', which contributes around 70% of the total area under pomegranate cultivation followed, by Karnataka and Andhra Pradesh. In Maharashtra, production of pomegranate is mainly concentrated in the western Maharashtra and Marathwada region. Pomegranates are commercially cultivated in Solapur, Sangli, Nashik, Ahmednagar, Pune, Dhule, Aurangabad, Satara, Osmanabad and Latur districts. The variety Bhagwa, extensively cultivated in Maharashtra is suitable for export purpose.

Export Potential

India has great potential for export of pomegranate fruit considering the better keeping quality and availability of fruits throughout the year. The commercial varieties, Bhagwa, Ganesh and Arakta are suitable for export market. Some gulf countries like Saudi Arabia, Qatar, Kuwait and Bangladesh are the major importing countries of Indian pomegranate (Table 34.1.). During the year 1992-93, a total quantity of 17,903 metric tonnes of fruit worth at Rs. 215.60 lakh was exported from India. However, the export of pomegranate increased to 25,000 metric tonnes in the year 2006. Afterwards, the export has crossed the mark of 35,000 tonnes per annum.

Table 34.1: Export Specification for Pomegranate

Variety	Countries		
	Middle East	Netherlands/ Germany	U.K.
Bhagwa / Ganesh	300-450g/ Red	250-300g/Red	250-300g/Red
Arakta / Mridula	200-250g/Deep Red	200-250g/Deep Red	200-250g/Deep Red
Packing	5kg	3kg	3kg
Storage	5°C	5°C	5°C
Export	By sea	By sea	By sea

Supply of pomegranate from Spain is mostly during the month of October – December whereas pomegranate is available throughout the year in India. Pomegranates are available round the year in Maharashtra and Tamil Nadu states. This advantage of round the year availability has to be tapped for export promotion. To facilitate the export potential, the pomegranates are available in different period in various states / Union Territories of India as given below.

      Andaman & Nicobar   : November - January

      Andhra Pradesh   : March - April

      Chattisgarh   : June - August

      Dadra & Nagar Haveli   : March - May

Daman & Diu	:	March - May
Gujarat	:	March - May
Himachal Pradesh	:	August
Karnataka	:	July- December
Madhya Pradesh	:	June - July
Maharashtra	:	January- December
Nagaland	:	September
Odisha	:	November - February
Puducherry	:	April - May
Rajasthan	:	March, August, November
Sikkim	:	December - January
Tamil Nadu	:	January - December
Tripura	:	July- August

During 2012-13, in India, Maharashtra produced 54.8% of pomegranates followed by Karnataka (20.2%), Gujarat (10.6%), Andhra Pradesh (8.3%), Madhya Pradesh (3.1%), Tamil Nadu (1.6%), Rajasthan (0.7%) and others (0.7%).

Major Commercial Varieties

There are several types of edible pomegranate besides the ornamental types with double flowers, largely sterile, which are not grown for edible fruit. The cultivars of edible type pomegranate which are grown commercially around the world include Wonderful in California and Israel; Mollar and Tendral in Spain; Schahvar and Robab in Iran; Hicaznar and Beynar in Turkey as well as Zehri and Gabsi in Tunisia. There are more than 25 pomegranate varieties grown in different parts of India (Table 34.2). Fruits of the wild-type pomegranate are acidic in nature whereas cultivated type are sweet. Among the commercial cultivars, 'Bhagwa' occupies the major area under cultivation in India. Ganesh, Ruby, Arakta, Mridula, Jyoti, Dholka, Jalore Seedless, *etc.* are some of the other varieties cultivated on commercial scale.

Table 34.2: Commercial Varieties of Pomegranate Grown in India

State	Varieties
Andhra Pradesh	Bhagwa, Ganesh, Ruby
Gujarat	Dholka, Bhagwa
Himachal Pradesh	Bhagwa, Ganesh, Kandhari
Jammu & Kashmir	Bhagwa, Ganesh

Contd...

State	Varieties
Karnataka	Ruby, Bhagwa, Jyoti, Mridula
Madhya Pradesh	Bhagwa, Ganesh
Maharashtra	Bhagwa, Ganesh, Mridula, G-137, Arakta
Rajasthan	Jalore Seedless, Jodhpur Red
Tamil Nadu	Bhagwa, Ganesh

Post Harvest Management

Although pomegranate is not a highly perishable fruit, yet appropriate postharvest management plays a crucial role in retention of postharvest quality for fetching better return to this high value produce.

Maturity Indices

The fruits of pomegranate should be harvested before they become overripe and crack (split) open, especially under rainy conditions. Maturity indices vary from variety to variety. The pomegranate fruit reaches full maturity (ripeness) within 4.5 to 6 months after full bloom, depending on the variety and climatic conditions. The fruits of pomegranate are harvested when they reach a certain size and attain desired skin color. Other maturity indices include titrable acidity (0.4-0.8%) and soluble solids content (13-17°Brix). Each pomegranate variety requires a certain 'total soluble solids / titrable acidity ratio' at harvest.

The mature fruit gives a metallic sound when tapped. Properly mature fruits are easily scratched with finger nails. With the advancement of maturity, the percentage of arils increases while that of seed, rind and rind thickness decreases gradually. The TSS: acid ratio increases and attains its peak (25 to 45). The pomegranate fruits contain about 45-61% juice on the basis of whole fruit weight or 76-85.5% on the basis of aril weight.

Harvesting

Pomegranate has a tendency to split open the fully ripe fruits as it is the nature's way of seed release and dispersal. The practice of early harvest in order to mitigate fruit cracking at maturity stage is one of the reasons of poor quality pomegranate in India. Late harvesting of fruits on the other hand results in a physiological disorder known as internal breakdown. ie. the discoloration of affected arils occur inside the fruit. Hence, harvesting should be done at appropriate time. Hence, pomegranate fruits are harvested only after attaining maturity in the plant.

Fruits are generally clipped when they are ripe. Depending on the *Bahar* treatments, pomegranate comes to harvest in installments. Usually the harvest commences in December- January and extends upto June-July. Fruits ripening from April to June often get affected by sunscald and may also crack if rains intervene or irrigation is irregular. In Deccan plateau and adjoining regions, the harvesting period of pomegranate is as follows (Table 35.3).

Table 35.3: The Harvesting Period of Pomegranate in India

Name of Bahar	Period of Flowering	Period of harvesting
Ambe bahar	Jan-Feb	Jul-Aug
Mrig bahar	Jun-July	Dec-Jan
Hasth bahar	Sep-Oct	Mar-Apr

While carrying out the harvesting operation, the following points may be kept in mind. The mature fruits alone are harvested during morning or evening without pulling from branches as the fruits don't ripe after harvest even with ethylene treatment. The scissors / clippers for picking need to be sanitized using 1% sodium hypochlorite. The fruits are to be clipped at the base of stalk without inflicting any injury to the rind and keeping the crown intact. Ladders / tripod stands may be used to climb up and harvest the fruits from upright branches. Perforated plastic crates lined with 2mm foam at the base may be used to keep the harvested fruit in the field. The crates should be stacked under shade.

Curtailing the number of fruits per tree to its optimum level according to the age of trees upto 5 years and "thinning the excess fruits" helps to reap good quality fruits. Cluster bearing is a common phenomenon in pomegranate. The development of multiple fruits (two or more) often results in competition among fruits for fruit development, branch bending, inferior quality fruits *etc*. Hence, it is better to allow solitary fruit to develop at a point from leaf axil. Yield increases progressively and in the 10th year, a plant may produce 175-200 fruits per year. As overbearing of the plant often results in breaking of branches, optimum fruit load/ plant has to be maintained according to age of the plants. ie., 60 fruits/plant- 3rd year; 80 fruits/ plant-4th year; 100 fruits/ plant-5th year; 100-150 fruits / plant – beyond 5th year depending on the age, canopy, vigour *etc*.

Sorting and Grading

Pre-cooling of the fruits at farm level is done preferably in cool chamber or in the shade. Grading of pomegranate fruits is important to obtain reasonable price for the produce from export and local market. Cracked, split, diseased and borer infested fruits should be separated and removed. The fruits are graded on the basis of their weight, size, and external (rind) colour (Table 34.4). The pomegranate fruits are divided into different grades *viz.*, Super, King, Queen and Prince for export market besides 12A & 12 B grades.

Table 34.4: Grades in Pomegranate for Export Purpose

Grade	Characteristics
Super	Attractive red colour fruits, > 750g / fruit, free from spot in the rind
King	Attractive red colour fruits, 500-750g/ fruit
Queen	Bright red colour fruits, 400-500g/fruit, free from spot
Prince	Red colour fruits, fully ripe fruits, 300-400g/ fruit

The fruits are categorized into 12A & 12 B grades. The fruits weighing around 250-300g and free from spots are graded as 12A. The fruits weighing around 250-300g with few spots on the surface are graded as 12B. The fruits of 12A grade are generally preferred in southern and northern India over 12B. After grading, the fruits are sometimes treated with ethyl oleate to provide luster.

Packaging of Pomegranate

The size of packages for pomegranate changes according to their grade. Corrugated fibre board (CFB) boxes are used for packaging since they are light in weight, cause less or no damage to fruits and easy to handle. CFB cartons of a standard size (40 cm x 20 cm x 24 cm) are used for packaging. Four fruits of Super size or 6 fruits of King size are packed in each CFB box of size 13"x9"x4" lbh. Nine fruits of Queen size are packed/ CFB box (15"x11"x4" lbh). Twelve fruits of Prince size are packed / CFB box (14"x10"x4" lbh). The white coloured, 5 ply CFB boxes are generally preferred for export purpose, whereas red coloured 3 ply CFB boxes are used for domestic markets. The fruits are wrapped in tissue paper and arranged in 2 rows for export market. Paper shreds are generally used as cushioning material.

The graded fruits are placed on cushioning material followed by an attractive red colour paper on the boxes. Boxes made up of light wood, bamboo basket are also used for packaging. Dry grass, rice straws, or paper are used as cushioning material at the bottom and top of box or basket. Upto 6 dozen fruits are packed in each basket for transport by trucks.

Standards for Export of Pomegranate

There are some desirable fruit characters of fresh pomegranate which are considered indispensable for export purpose. These are:

- Dark rose pink colour of the fruit.
- Fruit weight around 250-450 g.
- Round and globose shape of the fruit.
- Uniform size and shape of the fruit in a pack or box.

- Dark rose pink arils.
- Softness of the seeds.
- Higher sugar content with soluble solids content around 16-17° Brix.
- Free from scars, disease spots, insect injury, scratches, *etc.*
- Smooth cutting at the stem end.
- Pleasant flavour and aroma.
- Intact bracts (calyx) without any damage.

In Europe and other countries, pomegranate varieties such as Bhagwa, Mridula *etc.*, have good preference. Soft seeded, coloured varieties with higher juice content and easy to remove arils are preferred. The fruits weighing more than 500 g with superior qualities have immediate and ready acceptance in the international market.

35

Entrepreneurship Developement in Pomegranate through Value Addition

Nilesh N. Gaikwad, Ram Krishna Pal and K. Dhinesh Babu
ICAR-National Research Centre on Pomegranate, Solapur-413255, Maharashtra
E-mail : nileshgaikwad98@gmail.com

Introduction

Pomegranate fruit has wide acceptability among the consumers because its arils have attractive colour, sweet acidic taste, refreshing juice and is known for its nutraceutical value. India is one of the leading producers of the pomegranate in the world. In India area under pomegranate cultivation is 1.93 Lakh ha with production of 2.19 lakh tonnes in the year 2015-16 (NHB 2016). The area under cultivation of pomegranate has been increased by 70.80 % during last five years from 1.13 Lakh ha in year 2012-13 to 2.19 Lakh ha in year 2015-16. The pomegranate is grown in the states of Maharashtra, Karnataka, Gujarat, Andhra Pradesh, Himachal Pradesh, Rajasthan and Tamilnadu. The Export of pomegranate from India in year 2015-16 was valued at Rs 4160 million. The expected rise in production of pomegranate has compelled to rethink about future marketing and utilization strategy for this high value produce. In this context the post-harvest management of produce to improve its shelf life by adopting modern handling, storage, packaging and transportation practices is of high importance for marketing fruit to distant places in India and international destinations. The pomegranate can be processed in to various value added products as well as processing industry wastes/by products can be utilized for high value nutraceutical and pharmaceutical products.

Nutrients and Phytochemical

The fruit is rich in flavonoids, anthocyanins, punicic acid, ellagitannins, alkaloids, fructose, sucrose, glucose, simple organic acids, and other components (Aida Zarfeshany*et al.et al.* 2014). The pomegranate fruit contains approximately 40 % arils, 10 % seeds and 50 % peel. Pomegranate arils contain 85% water, 10% total sugars, mainly fructose and glucose, and 1.5% pectin, organic acid,such as ascorbic acid, citric acid, and malic acid, and bio-active compounds such as phenolics and flavonoids, principally anthocyanins (M. Viuda-Martos, 2010). Pomegranate arils provide 12% of the Daily Value (DV) for vitamin C and 16% DV for vitamin- K per 100g serving. Very recently it has been found as a rich source of Fe and Zn. The red color of juice can be attributed to anthocyanins, such as delphinidin, cyanidin and pelargonidin glycosides. The pomegranate peel is an important source of bioactive compounds such as phenolics, flavonoids, ellagitannins, and proanthocyanidin compounds, minerals, mainly potassium, nitrogen, calcium, phosphorus, magnesium, and sodium, and complex polysaccharides. Pomegranate seeds are approximately 10 per cent of the fruit weight. Pomegranate seeds are excellent sources of dietary fiber. The soft seeded varieties of pomegranate contain seeds oil to the tune of 25-26 % (V/W). The pomegranate seed oil contains more than 70 % of conjugated linolenic acid. Pomegranate seed oil contains punicic acid (65.3%), palmitic acid (4.8%), stearic acid (2.3%), oleic acid (6.3%) and linoleic acid (6.6%).

Potential Health Benefits of Pomegranate

The extracts from different parts of pomegranate had shown therapeutic properties that targets wide range of diseases including cancer, cardiovascular disorders, reproductive disorders including male infertility, skin diseases, dental problems, stomach disorders,Alzheimer's disease (Jurenka, 2008), aging, and AIDS (Neurath, *et al.* 2010). It also has anti-inflamatory properties.

Table 35.1: Principal constituents of different parts of pomegranate tree and fruit.

Plant component	Constituents
Pomegranate juice	Anthocyanins, glucose,organic acid, ascorbic acid, EA, ETs, gallicacid, caffeic acid,catechin, quercetin, rutin, minerals
Pomegranate seed oil	Conjugated linolenic acid, linoleic acid,oleic acid, stearic acid,punicic acid,eleostearic acid,catalpic acid
Pomegranate peel	Luteolin, quercetin,kaempferol, gallagic,EA glycosides, EA,punicalagin,punicalin,pedunculagin

Contd...

Plant component	Constituents
Pomegranate leaves	EA; fatty acids
Pomegranate flower	Polyphenols, punicalagin, punicalin, EA
Pomegranate roots and bark	Alkaloids, ETs

Michael Avirama*et al.et al.*(2004) investigated the effects of pomegranate juice (PJ) consumption by atherosclerotic patients with carotidartery stenosis (CAS) on the progression of carotid lesions and changes in oxidative stress and blood pressure. The studies revealed that pomegranate juice (PJ) consumption by patients with carotid artery stenosis (CAS) possesses anti-atherosclerotic properties as it significantly reduced common carotid IMT in association with a decrease in systolic blood pressure, and a substantial inhibition of lipids peroxidation in serum and in LDL.

Research on breast cancer cell lines demonstrated that pomegranate constituents efficiently inhibited angiogenesis (Toi*et al.*, 2003), invasiveness (Kim*et al.*, 2002), growth (Mehta and Lansky, 2004), and induced apoptosis (Jeune*et al.*, 2005). Its anti-invasive, anti-proliferative, and anti-metastatic effects were attributed to the modulation of Bcl-2 proteins, upregulation of p27 and p21, and down regulation of cyclin-cdk network (Faria and Conceic 2011).

In research conducted on pomegranate seed oil in skin cancer (Hora*et al.* 2003,Sayed*et al.* 2006) demonstrated that pomegranate oil has chemopreventive efficacy in mice. Reduced tumor incidence (7%), decrease in tumor numbers, reduction in ornithine decarboxylase (ODC) activity (17%), significant inhibition in elevated tissue plasminogen activator (TPA)-mediated skin edema and hyperplasia, protein expression of ODC, COX-2 and epidermal ODC activity have been reported with pomegranate oil treatments.

Inflammation can cause various physical dysfunctions. Pomegranate is widely used as an antipyretic analgesic in Chinese culture. Pomegranate has shown potential nitric oxide (NO) inhibition in LPS-induced RAW 264.7 macrophage cells (Chia-Jung Lee*et al.*, 2010).It is demonstrated that pomegranate fruit extract has an embryonic protective nature against adrianycin-induced oxidative stress, where adrianycin is a chemotherapeutic drug used in cancer treatment (Kishore, 2009). Moreover, pomegranate juice consumption could increase epididymal sperm concentration, motility, spermatogenic cell density, diameter of seminiferous tubules and germinal cell layer thickness (Turk *et al.,* 2008).

Pomegranate Processing and Value Addition

Large quantities of fruits are consumed in fresh form for table purpose. The products like pomegranate juice, ready to serve (RTS) beverage, squash, jelly and *anardana* (dried arils/seeds) are also being prepared on commercial scale worldwide. Apart from its demand for fresh fruits and traditional processed products, the innovative and non-traditional high value processed products like pomegranate wine, tea, minimally processed arils, and seed oil are also gaining importance in world trade.

Pomegranate Aril Extraction

The processing of pomegranate for value addition requires the separation of arils from the fruits. The process is very tedious and time consuming when done manually. The mechanized machines and hand tools have been developed for the separation of arils from the fruit. The separated arils are then used for further processing in to the various value added products. The CIPHET an ICAR institute from India has developed two types of aril extractors shown in the figure.One is a hand tool for breaking of pomegranate and consequently easy separation of arils from its peel and another is motorized aril extractor of about 500 kg/h capacity. The separated arils can be packaged in punnets for further sale or converted in to *anardana,* juice *etc.*

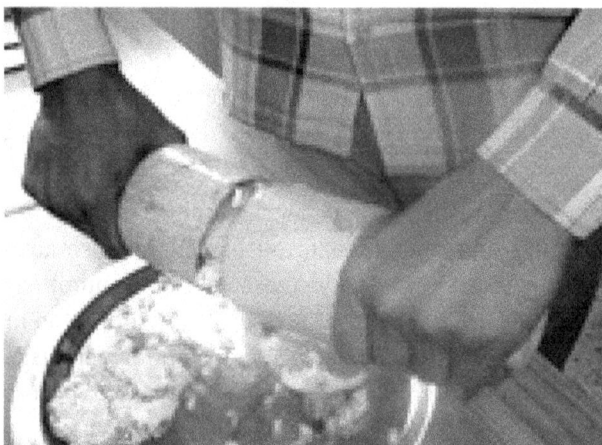

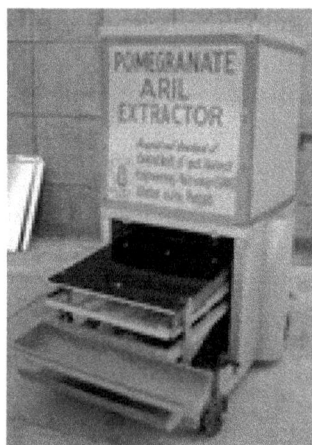

Hand Tool (Left)) and Motorized Versions (Right) of CIPHET Pomegranate Aril Extractor

Minimally processed pomegranate arils

The pomegranate is consumed mainly fresh, but the difficulty encountered in separating the edible arils from the fruit has several limitations for its direct consumption unlike the other fruits e.g. oranges, banana, grapes *etc.* Hence, minimal processing of pomegranate is of great importance for convenience of the consumers. Commercialization of minimally processed and "ready-to-eat" fresh arils is the good alternative. In the minimal processing generally various hurdles are created for spoilage of freshly extracted arils using low temperature, pH regulation and GRAS (generally recommended as safe) chemicals as anti-microbial agents. The minimal processing consists of washing with sanitizing agents- to reduce the primary inoculum load, pH modification, use of antioxidant agents, temperature control and others, to control partially the high perishability of the fruits. On the other hand, the selectively permeable polymeric films for packaging of minimally processed pomegranate arils are used for generation of Modified Atmospheric Packaging (MAP) system in order to develop a micro controlled environment that reduces the respiratory activity and maintains unfavorable conditions for the action of many contaminating microorganisms. Researches carried out on minimal processing of pomegranate arils show browning produced by the oxidation of the phenolic compounds during the storage, indicating that the stabilization of anthocyanin pigments is essential in order to achieve a good quality. Hess-Pierce and Kader (1997) showed that pomegranate arils of variety Wonderful can be stored up to 16 days at 5°C with 20% CO_2, without changes in the physical and chemical characteristics.

Pomegranate Juice

Pomegranate juice is nutritionally an important beverage and it is consumed frequently for its phenolic compounds (such as anthocyanins, ellagic acid, phytoestrogenic flavonoids and tannins).Pomegranate juices have shown an antioxidant capacity three times higher than red wine and green tea. The screw presses are used prominently for juice extraction from pomegranate arils in small scale processing. The processing of pomegranate juice at commercial level involves squeezing juice from the fruit with the seeds and the peels together. The juice extraction from whole fruits is carried out by subjecting the cut fruits to hydraulic press. The pressure of less than 100 psi is used to avoid undue yield of tannin. The juice from crushed whole fruits contains excess tannin from the rind (as much as 0.175%) and which is precipitated out by a gelatin process.

The pomegranate juice is susceptible to microbial contamination with acid-tolerant bacteria, fungi (yeasts and moulds), and pathogenic bacteria, that leads to the deterioration of nutritional and sensorial properties such as functional ingredients, colour, flavor, and odor, as well as food-borne diseases due to the pathogenic bacteria or toxigenic fungi.

The pomegranate juice is clarified by heating in a flash pasteurizer at 79-82 °C, cooling, settling for 24 hours racking up and filtering or decanting. The clear juice can be preserved by heat treatment or by using chemicals (1000 ppm sodium benzoate). After heating at 80°C it is filled in to bottles while still hot.The glass bottles are crown corked and further processed for better storability of juice. The useof PET bottles for packaging of juices are preferred now a days.Thermal pasteurization techniques decrease the nutritional and organoleptic quality and freshness of juice. Some of the non-thermal technologies having potential to replace thermal processing of foods include membrane filtration, osmotic dehydration, pulse electric field, ultrasound, irradiation, high pressure, active packaging and ozone treatment.

Ready To Serve Beverage from Pomegranate

The pomegranate RTS beverage is developed by NRC Pomegranate, Solapur with 20% of juice and adjusting TSS and acidity to 15°B and 0.375% by addition of sugar and citric acid. The RTS developed is pasturized at 80°C for 2 minutes. The beverage can be packaged in glass, PET or PP bottles by filling it when it is still hot. The RTS developed had higher acceptability on the basis of sensory score. Carbonated RTS beverage can also be prepared with 15 percent juice level and 12 percent sugar with 0.3 percent acidity level with carbonation at 80 psi of CO_2 pressure.

Pomegranate Jelly and Jam

Pomegranate jellyisprepared from combination of juice sugar 1:1 and citric acid as acidulantis of better quality, colour, flavour and acceptance. Similarly, pomegranate jam is made by concentrating pomegranate juice, heating the mixture on slow fire for a long period until the consistency of finished product isthick and TSS68-70°Brix. It can be stored for one year.

Anardana

The conventional utilization of wild pomegranate fruits is mainly in the form of dried arils, which is called pomegranate raisins or '*Anardana*'.Anardana has a distinct tart flavour, and is commercially available in many Asian countries, where it is consumed in large quantities. Anardana is also used in the ayurvedic medicine as digestive and stomachic. The cracked fruits at matured stage can be economically used traditionally for preparation of dried value added product called *anardana*. It is used as acidulent and condiment in Indian culinary or traditional system of medicine. The cabinet drying at 55°C for 7 hours of the arils is best for getting quality *anardana*. Varieties having high natural acidity are suitable for preparation of anardana. Development of few varieties / hybrids for production of good quality *anardana* at NRCP is in progress.

Pomegranate Seed Oil

Pomegranate seed oil is a unique natural product, with no match among all other plant oils. It is one of only about six plant sources known that contain conjugated fatty acids. Conjugated fatty acids are important because they inhibit eicosanoid metabolism at several points in the synthesis of prostaglandins from arachidonic acid. This makes them significant natural anti-inflammatory agents. Pomegranate seed oil contains other significant bioactive compounds. It is the richest known plant source of a steroidal estrogen, estrone. Studies have shown that the oil contains over 3 mg/g of 17-alpha-estradiol, the mildest and safest steroidal estrogen and an exceptionally potent antioxidant and brain-preserving compound. Other important compounds found in the Oil include gamma-tocopherol, a rare and potent form of Vitamin E and the phytosterols: beta-sitosterol, stigmasterol and campesterol. It has been linked to improve heart health and also known to protect against cancer (Hernandez, *et al.*, 2002; Lansky and Newman, 2006) and artheroschlerosis (Boussetta, *et al.*, 2009). The studies on determination of pomegranate seed oil content with soxhlet method and using petroleum benzene a non-polar solvent were carried out. The average per cent oil in Bhagwa, Ganesh and Araktacultivars was found to be 28%, 26.43 % and 23.70% (w/w) and average specific gravity of pomegranate seed oil 0.90, 0.92 and 0.92 respectively. The fatty acid profile of Bhagwa and Ganesh cultivar with HPLC revealed both prominent cultivars grown in India have high level of linolenic acid content.

Fatty acid profile	Bhagwa Seeds	Ganesh Seeds
Linolenic (%)	68.7	69.8
linoleic (%)	13.9	11.4
Oleic (%)	12.0	9.1
Palmitic (%)	2.1	2.4
Stearic (%)	1.9	2.1

The pomegranate seed oil is mainly extracted through cold pressing due to its high value and tendency for oxidation. Large amounts of conjugated unsaturated fatty acid in its triglyceride composition which are very sensitive to heat and easily undergo cis/trans isomerization forbids the use of more effective thermal pressing or microwave-assisted extraction. The protocol of seed oil extraction involves seed separation from pomegranate marc (left out portion after juice extraction). Followed by, washing of seeds for the removal of remaining pulp from seed surface. Drying of seeds at 40°C in tray dryer for 24 hrs and grinding of seeds in a grinder for size reduction. The ground seeds can then be used for oil extraction in the cold press at temperature of 40°C. The cold pressing of ground Bhagwa seeds in hydraulic type of cold press at pressure of 5 MPa and temperature of 40°C could yield 12.50 % of oil on w/w basis.

Extraction of Antioxidants from Pomegranate Marc

Recently, natural antioxidants have become very popular for medical and food applications and are preferred by consumers than synthetic antioxidants, such as BHA and BHT (Wang Zhenbin*et al.*, 2011). The commercial processing of pomegranate for juice involves squeezing juice from the fruit with the seeds and the peels together. The resulting marc on a weight basis consists of approximately 73 % peels and 27% seeds and has a high potential for value addition as a source of phenolic, pro -anthocyanidin and flavonoids which are herein referred to as antioxidants. The pomegranate peel in particular possesses relatively higher antioxidant activity than seed and pulp. This is because of polyphenols in peel composed of condensed tannins and anthocyanin.

Bio-colours

The rind of pomegranate contains a considerable amount of tannin, about 19% with pelletierine (Adeel*et al.*, 2009; Tiwari*et al.*, 2010). The main coloring agent in the pomegranate peel is granatonine which is present in the alkaloid form N-methyl granatonine(Goodarzian and Ekrami, 2010). This compound gives colour from pomegranate peel. The natural dyes from peel of pomegranate are used in colouration of lipsticks and other cosmetics. They are produced by water extraction. The colouring matter can be extracted with super critical fluid extraction method.

Pomegranate Wine

Pomegranate wine is the product of anaerobic fermentation by yeast in which the sugars are converted into alcohol and carbon dioxide. Pomegranate wines have higher antioxidant activity and total phenol content than the red wine. Melatonin (N-acetyl-5-methoxytryptamine) is a neuro-hormone related to a broad array of physiological functions and proven therapeutic properties.Melatonin was observed to be absent in pomegranates juices but it is detected in prominent amounts with respect to other food matrixes (0.54–5.50 ng/mL) in pomegranate wine. The presence of this biogenic amine makes pomegranate wine use as bioactive phytochemicals supplementation.

The protocol for preparation of pomegranate wine was refined at NRCP, Solapurwith the inclusion of enzyme pre-treatment. Pomegranate wine was prepared from pomegranate juice using the shake flask culture method by adopting the following steps. Pomegranate juice was extracted by cutting fruits of cv. Ganesh in to two halves and pressing in a hand press. The TSS was measured with refractometer and then adjusted by adding sugar. Potassium meta-bisulphite is added to restrict the growth of undesirable microorganisms. The juice was pasteurized and treated with enzyme. Fermentation of juice was carried out using yeast in incubator shaker at 20°C for about10-12 days. Wine was then clarification by centrifugation. The bentonite is also added to wine. The wine is flash pasteurized at 60°C, bottled hot and sealed. The wine prepared by the above method was found to be superior in clarity / transparency compared to the control.

Conclusion

Post-harvest management and value addition plays a crucial role to sustain the profitability of pomegranate production with t rapid rate of expansion in pomegranate area and production. Research on development of need based and cost-effective post-harvest management practices for maintenance of post-harvest quality during handling will go a long way in expansion of market destination for domestic as well as export trade. Similarly, development of value added products from pomegranate will facilitate the total utilization of this high value commodity generating employment for rural youth.

36

Utilisation of Pomegranate by Products for Dyeing of Cotton Textiles

Sujata Saxena[1], R.R. Mahangade[1], R.S. Narkar[1] and Ram Chandra[2]

[1]ICAR-Central Institute for Research on Cotton Technology,
Mumbai-400 019, Maharashtra
[2]ICAR-National Research Centre on Pomegranate, Solapur-413 255, Maharashtra
E-mail: saxenasujata@rediffmail.com

Introduction

As pomegranate plant parts are considered to be rich in tannins, these materials can be used as a tannin source for mordanting of cotton in natural dyeing and also as a natural dye. Today, textiles dyed with natural dyes are preferred by environment conscious consumers due to their eco-friendly nature and fetch higher prices. In this study fruit rinds and flowers of popular varieties of pomegranate were analysed for their tannin content and their potential for use as dye as well as mordant for dyeing of cotton was explored. Colour parameters and colour fastness properties of dyed cotton fabrics were evaluated. Results were promising and the same have been discussed in this paper. Utilisation of these by-products of pomegranate cultivation for natural dyeing can thus result in good value addition.

Materials and Methods

Plant materials used in this study were collected from NRCP, Solapur farms and comprised of rinds and flowers of cultivars Ganesh and Bhagwa and flowers of cultivars Arakta and Mridula. These were dried in shade and powdered in a mixer. Harda (*Terminalia chebula*) powder, used commonly as tannin mordant in natural dyeing was obtained from M/S Natural Dye Resources, Sawantwadi.

Fabric- Commercially scoured and bleached 100% cotton plain weave poplin fabric procured from a local mill was used for this study.

Analysis of Tannin Content- All plant materials were analysed for tannin content by modified AOAC method using Folin- Denis reagent. Plant material weighing 250 mg was suspended in 50 ml distilled water and kept on an orbital shaker for 30 min. It was then filtered through G-1 glass sintered crucible and total volume of the filtrate was made to 100 ml in a volumetric flask. One ml of this solution was added to 75 ml distilled water kept in another 100 ml volumetric flask. Five ml each of Folin- Denis reagent and 35% sodium carbonate solution were added to the flask and the volume was made upto the mark. After allowing the colour to develop for 30 min, absorbance of the solution was measured on Analytic Jena UV- visible spectrophotometer at 760 nm. Tannin percentage was determined from the standard curve prepared with standard tannic acid solutions.

Treatment of Cotton Fabric with Plant Materials (Mordanting)- Fabric swatches weighing 6g were treated with the aqueous dispersion of the powdered plant materials (20% on weight of fabric) in glass conical flasks kept on an orbital shaker at room temperature (27⁰C) for 60 min. Material to liquor ratio of 1: 25 was used. Fabrics were then removed, treated with potash alum solution (20% o.w.f., neutralized with sodium hydroxide) for 30 min for fixing the tannins, washed well with water and air dried.

Evaluation of Treated (Mordanted) Fabrics- A yellow colour was developed on cotton fabrics for all plant materials. Colour parameters of these mordanted samples were determined on Spectrascan 5100 computerised colour matching system. Colorfastness properties of treated materials were determined as per the following standard test methods-

1. Colourfastness to light - IS 2454 : 1985 Reaffirmed : 2006
2. Colourfastness to washing: IS/ISO 105 - C 10: 1985 Reaffirmed: 2006 Test No. B (2)
3. Colourfastness to perspiration - ISO - 971 : 1983 Reaffirmed : 2004

Dyeing of Treated (Mordanted) Fabrics- All above mordanted fabric samples were further dyed with natural red dye manjith (*Rubia cordifolia*) to 20% shade (on weight of fabric) in a launderometer. It is a mordant dye and needs a mordant for fixation onto cotton. Dye extract was prepared by boiling the required quantity of powdered manjith dye with water and filtering the solution. Material to liquor ratio was again kept 1: 25. Dyeing started at room temperature and the temperature was gradually increased to near boil and dyeing was continued at this temperature for 30 mins. Dyed samples were then removed from dye bath, rinsed with water and soaped with non-ionic detergent solution (2g/ L) at 60°C for 10 min, rinsed again with water and air dried.

Dyed fabrics were also evaluated for various colour parameters and fastness properties as described above.

Results and Discussion

Tannin content of the plant materials is presented in table 36.1 and it is seen that all tested pomegranate plant materials are rich in tannins with their tannin content ranging from 14-36%. Highest tannin content of 36% is seen in cv. Ganesh rind which is quite close to the tannin content of Harda fruits. This cv. seems to be rich in tannin as among the four flower cultivars studied, flowers of this variety had the highest tannin content. Tannin content of Bhagwa rinds was lower (20%) and similar to the tannin content observed for Bhagwa flowers. Flowers of Arakta and Mridula variety had lower tannin content.

Table 36.1: Total Tannin Content of Powdered Plant Materials

Sr. No.	Plant Material	Percent Total tannin
1.	Pomegranate rind cv. Bhagwa	20
2.	Pomegranate rind cv. Ganesh	36
3.	Pomegranate Flower cv. Bhagwa	20
4.	Pomegranate Flower cv. Ganesh	22
5.	Pomegranate Flower cv. Arakta	14
6.	Pomegranate Flower cv. Mridula	15
7.	Dried Harda Fruits	50

Cotton fabric developed yellow colour upon treatment with these plant materials which suggests that these materials, in addition to tannin also contain a yellow dye which was also indicated by the uv- visible spectra of these extracts. There was good absorption observed throughout the uv region but in visible region highest absorbance was observed at 400 nm which corresponds to the yellow region. Colour parameters of the treated cotton fabrics are tabulated in table 36.2. It is seen that highest colour strength (K/S value) was observed in case of Ganesh rind which was higher than that obtained for Harda fruits. Also this fabric had highest positive value of b* indicating good yellow colour of the fabric. This indicates good presence of yellow colourant in this plant material along with high tannin content. Tannin alone does not result in yellow coloration as the sample treated with pure tannic acid had very low colour strength. Cotton treated with other pomegranate materials showed good yellow colour (positive b* value) but its depth was less as seen from lower colour strength values.

Table 36.2: Colour Parameters of Mordanted Cotton Fabric Samples

Sr. No.	Tannin Mordant used	K/S at λ max	λ max (nm)	Colour coordinates (CIE Lab 1976)				
				L*	a*	b*	C*	H*
1.	Tannic acid	0.44	420	83.49	0.95	13.74	13.77	85.99
2.	Harda	4.53	420	73.20	0.33	35.90	35.90	89.44
3.	Pomegranate rind cv. Bhagwa	2.67	420	73.80	2.39	30.60	30.69	85.51
4.	Pomegranate rind cv. Ganesh	5.27	420	71.84	1.96	38.86	38.92	87.14
5.	Pomegranate flower cv. Bhagwa	2.15	420	75.36	2.30	29.70	29.78	85.53
6.	Pomegranate flower cv. Ganesh	2.40	420	74.68	2.72	30.69	30.81	84.91
7.	Pomegranate flower cv. Arakta	1.96	420	75.64	2.47	28.54	28.54	85.00
8.	Pomegranate flower cv. Mridula	1.76	420	76.26	2.17	27.30	27.39	85.42

Fastness of the colour of a dyed fabric to the action of various agencies to which a fabric is exposed during its use such as light, washing, perspiration *etc.* is a very important criteria to determine the suitability of a dye for textile applications. In order to determine performance of the colorant present in pomegranate parts, cotton fabrics dyed with pomegranate rinds were evaluated for various colourfastness properties and the results are summarized in table 36.3. Colour fastness to washing of pomegranate rind treated samples was good (4 on a scale of 1-5 with 5 being the best) and the colour became little darker upon washing and same was observed for Harda also. At the same time, there was no staining on the adjacent white cotton and wool samples. Colourfastness to light of Ganesh rind treated sample was very good (>5, no fading under test conditions) and was at par with harda treated sample. Performance of Bhagwa rind treated sample was little lower but could still be classified as good. Perhaps lower tannin content in Bhagwa has resulted in a bit lower fastness rating. Colourfastness of all samples to both acidic and alkaline perspiration ranged from good to satisfactory. Again, Ganesh rind treated samples performed better and were at par with harda.

Table 36.3: Colourfastness Properties of Mordanted Samples

Tannin Mordant used	CF to Washing			CF To Light	CF to acidic perspiration			CF to alkaline perspiration		
	Colour change	C	W		Colour change	Staining		Colour change	Staining	
						C	W		C	W
Harda	4 d	4-5	5	>5	4	4-5	4-5	4-5	4-5	4-5
Pomegranate rind Bhagwa	4 d	4-5	5	4	3	4-5	4-5	4	4-5	4-5
Pomegranate rind Ganesh	4d	4-5	5	> 5	4	4-5	4-5	4-5	4-5	4-5

CF - Colourfastness, C-Cotton, W - Wool, d – darker

Thus these pomegranate by products are suitable for use as yellow dye. These treated samples can also be considered as mordanted samples and can be further dyed with other natural dyes. In the present study these were dyed with manjith which is a red natural dye and needs a mordant for fixation onto cotton as colour yield on unmordanted material is very low. Good red colour was obtained for all samples and colour parameters of dyed samples are tabulated in table 36.4 below.

Table 36.4: Colour Parameters of Mordanted and Manjth dyed cotton samples

Sr. No	Tannin Mordant used	K/S at λ max	λ max (nm)	Colour coordinate (CIE Lab 1976)				
				L*	a*	b*	C*	H*
1.	Tannic acid	5.76	510	46.06	35.59	17.17	37.31	17.42
2.	Harda	6.96	510	46.50	32.40	22.21	39.25	34.42
3.	Pomegranate rind cv. Bhagwa	4.79	510	47.26	32.54	18.90	37.62	30.12
4.	Pomegranate rind cv. Ganesh	8.32	510	44.88	32.91	23.74	40.58	35.76
5.	Pomegranate Flower cv. Bhagwa	4.47	510	49.31	32.28	14.96	35.58	24.86
6.	Pomegranate flower cv. Ganesh	4.69	510	48.82	32.45	15.83	36.10	25.99
7.	Pomegranate flower cv. Arakta	3.66	510	51.61	30.93	13.78	33.86	23.99
8.	Pomegranate flower cv. Mridula	3.87	510	51.16	31.87	13.56	34.63	23.05

It is seen from the table 36.4 that Ganesh rind treated samples have developed darkest colour as their colour strength is highest and L*(lightness) value is lowest. It also had the highest total colour value (C*) and its performance was superior to even harda used traditionally as mordant.

Table 36.5: Colourfastness Properties of Mordanted and manjith dyed Samples

Tannin Mordant used	CF to Washing			CF To Light	CF to acidic perspiration			CF to alkaline perspiration		
	Colour change	C	W		Colour change	Staining		Colour change	Staining	
						C	W		C	W
Harda	4-5	4-5	4-5	4	4	4-5	4	4-5	4-5	4-5
Pomegranate rind cv. Bhagwa	4-5	4-5	4-5	3-4	3-4	4-5	4	4	4-5	4-5
Pomegranate rind cv. Ganesh	4	4-5	4-5	4	4	4-5	4	4-5	4-5	4-5

CF - Colourfastness, C-Cotton, W – Wool

Mordants not only provide site for attachment of natural dyes but also affect their fastness properties. Table 36.5 contains the colorfastness properties of harda and pomegranate rind mordanted and manjith dyed samples. All fastness properties studied range from good to very good and in that respect performance of pomegranate rinds as mordant is comparable to Harda. Hence pomegranate rinds can very well be used as mordant in natural dyeing of cotton.

Conclusion

It can therefore be concluded that pomegranate rinds and pomegranate flowers of the studied cultivars are rich in tannins and also contain a yellow colourant which in itself can be used as a dye with good fastness properties. These can also be used as mordant for cotton material in place of Harda for further dyeing with other natural dyes with good performance as was observed in the current study for dyeing with manjith.

Acknowledgement

Authors thank Director, CIRCOT for providing all facilities and permission to present this paper.

37

Agrochemical Residues in Pomegranate: Issues and Strategic Solutions

Kaushik Banerjee, Ahammed Shabeer and T.P. Sagar Utture

ICAR-National Research Centre for Grapes, Pune-412 307, Maharashtra
E-mail: kbgrape@yahoo.com

Introduction

Pomegranate (*Punica granatum* L.) is one of the important fruit crops of tropical and subtropical regions. In India, it is mainly cultivated in the states of Maharashtra and Karnataka. Due to high nutritional value its demand in the local and international markets is increasing day by day. It is known to have multiple health benefits and is therefore consumed in fresh forms as well as processed products such as juices, jams, *anardana, etc.*

Of the several horticulture products in India, the global demand for pomegranate has been increasing at a much faster rate compared to others. Pomegranate is currently ranked 18th in terms of fruits consumed in the world. It is thought that as a result of its health benefits, it is expected to move to 10th place in next 10 years (http://inifarms.com/market.html). But, the volume of export from India is only approximately 6-7% of the total pomegranate production. Indian horticulture always encounters the attack of pests and diseases, so is for pomegranate, which hampers the quality of the product. The main diseases attacking pomegranate include bacterial leaf and nodal blight, anthracnose, leaf spot and fruit spot, fruit rot, wilt complex, *etc.* Besides these diseases, the crop is attacked by various insect pests which include thrips, pomegranate butterfly, shoot hole borer, pomegranate aphid, mealy bug, leaf eating caterpillar, *etc.* The pomegranate crop is also susceptible to some disorders viz., internal breakdown of arils and sun scald, *etc.* Sucking pests like mealy bugs, thrips, *etc.* are the major obstacles during pomegranate cultivation. In such scenario, the application of pesticides is inevitable to control the myriad of the diseases and pests prevailing in tropical climate. Thus, farmers have no other choice but to apply pesticides of different chemical classes extensively for control of pests at various stages of pomegranate cultivation to increase the production and to keep the quality of produce intact.

On the other hand, considering the toxicity of the pesticides, public awareness over pesticide residues in food has increased steadily during the past 20-25 years. Hence most of the countries such as India, USA, European Union (EU), and other countries put in place stringent regulations (e.g. Food Safety and Standards Regulation of India) for pesticide residues in food, which includes setting up of maximum residue limits (MRLs). Due to the unavailability of scientific studies, in most of the commodities, the MRLs are set at the lowest limit as a consensus (limit of quantification) figure rather than an estimated value based on risk assessment studies. To avoid contradictions, such default MRLs are set at lower limit of analytical determination without any consideration of food safety or toxicological implications. Pomegranate is one of the important candidates of this group. Thus, in most of the cases any detection of pesticides in exportable pomegranate results in rejection of the consignments, and incurs financial loss to the farming community. On the other hand, in India also the Food Safety Standards Authority (FSSAI) is bring the concept of setting up of default MRL at 0.01 mg/kg in absence of risk assessment information and hence similar restrictions will be soon applicable for domestic food testing as well. Hence, monitoring of pesticide residues for both domestic consumption and export has become a priority in order to ensure consumer safety and meet the food safety/ quality standards.

Agrochemicals and Agrochemical Residues

Agrochemicals (majorly pesticides in the context of food safety) are the substances meant for management of various pests and diseases attacking the agricultural crops. In general, pesticides are chemical or biological agents that inhibit the activity of pests. The use of pesticides in agriculture helped to reduce the damage to the crops and significantly increased the crop productivity to a great extent. Although pesticides have huge benefits, the unregulated or uncontrolled use of it can create serious problems like potential toxicity to animal kingdom.

Pesticide residues are the remaining traces of substances resulted from the application of pesticide such as the parent molecule itself or its derivatives, degradation and conversion products, metabolites, reaction products and impurities which are considered to be of toxicological significance. Pesticide residues have been detected in various fruits, vegetables and other horticultural products and the persistent one can be magnified through the food chain and reach to the next tropic level.

The amount of pesticide remaining on/in horticultural products largely depends on the nature of the chemical, its application dose, number of applications, *etc.* The major sources of pesticide contamination in fresh fruits and vegetables can be the uncontrolled and excess use of such chemicals during the crop life cycle. The strategies for reduction of pesticides in final produce needs be implemented.

Management Strategies for Pesticide Residues

Good Agricultural Practice (GAP): Good agricultural practices (choice of agrochemicals for spraying, spray schedule, disease management, measures for fruit development, *etc.*) is key for minimizing the pesticide residues and thus to qualify the produce with the regulatory (MRL) requirements.

Establishment of Quality Control and Inspection Systems are essential components of GAP to ensure production of safe food. The GAP includes monitoring programmes concerning the hazardous contaminants being used in the agriculture and ensure that the contamination level in fruits and vegetables do not exceed the limits prescribed by the regulatory bodies. The monitoring of agro-inputs for pesticides reduces chances of unexpected contaminants in food and facilitates implementation of GAP.

Good Agricultural Practice (GAP) is an approach that improves environmental, economic and social sustainability of on-farm production and results in safe and quality food and non-food agricultural products. A GAP approach can contribute concretely to implement sustainable agriculture and rural development while addressing the demand-side priorities of consumers and retailers, the supply-side priorities of producers and labourers, and those institutions and services that are bridging supply and demand. While a GAP approach may respond to the growing demands of increasingly globalized and integrated agricultural sectors, it is also very important for local and national markets.

Pre-harvest Intervals- a Tool for Residue Management: Pre-harvest intervals for most of the commodities are based on field studies of residue dissipation so that the final produce is free of residues or the residues are below their respective MRLs. This practice simultaneously ensures crop protection as well as food safety. However, in case of pomegranate such systematic studies are absent for the Indian climatic conditions and cultivation practices. In India, very few pesticides are recommended for the management of pests and diseases in pomegranate (CIB&RC, 2). Therefore, farmers opt for using pesticides either based on their individual experience developed over time, recommendation from fellow farmers or pesticide company representatives, without based on scientific risk assessment. In most cases, the farmers are unaware of the residual levels of pesticides in the final produce. Thus, such practices, based on non-scientific scheduling of pesticide sprays reduce the reliability of the farmers, exporter and consumers on the quality of the produce at harvest. There is ample scope and huge market for export of pomegranates to other developed countries leading to inflow of foreign exchange. However, compliance to importing countries food safety requirements is important aspect. To fulfil the food safety requirements, it is therefore essential to manage the application of agrochemicals during cultivation so that the harvested produce is below the prescribed MRLs and complies with the quality grading requirements at the same time. Hence, there is an urgent requirement to prescribe waiting periods or pre-harvest intervals for pesticides to generate the desired produce that is fit for export as well as domestic consumption. The prevalent knowledge gaps of the PHIs could be addressed by conducting supervised field residue trials for agro-chemicals to be recommended for use in pomegranates. The dissipation of pesticides is studied in pomegranate for few insecticides (Utture et al, 2012), fungicides (Utture et al, 2011) and antibiotic residues (Jadhav et al, 2013). The dissipation of all the three fungicides was faster initially, and slowed down with the passage of time indicating a non-linear pattern of degradation. Out of the test fungicides, the dissipation of azoxystrobin was much faster than the other two chemicals for which the PHI value was found to be 9 days for standard dose and 20.5 days at the double dose. Since the dissipation of carbendazim in later stages was quite slower, the PHI appeared much longer (60 and 100 days at standard and double dose, respectively) despite the faster dissipation at initial days. The PHI for another fungicide difenoconazole was found to be 26 and 60 days at standard and double dose, respectively.In case of insecticides, for imidacloprid, the initial residue deposits (0.120 and 0.247 mg/kg for standard and double dose, respectively) were below the MRL (applicable for European Union). Hence, PHI does not apply for this chemical. The PHIs of the buprofezin, and dimethoate were found to be 10.5 and 31.5 days at the standard dose of application while at double the application dose the PHIs were 32.0 and 43.0 days, respectively.

Since only few pesticides are having PHIs available based on the supervised field trials, more and more such studies are required on various agrochemicals in pomegranate. This residue trial data and PHI will help to register more chemicals having label claim in pomegranate and farmers will be having wider choice of chemicals for diseases and pests management.

Translocation of agrochemical residues in different parts of produce: In modern day agriculture, large numbers of pesticides with different mode of action (systemic, surface active, translaminar, *etc.*) are applied on the crops. There exists significant deficiency in information regarding the extent of translocation of these residues to different parts of the edible (arils) as well as the non-edible part of the pomegranate fruit (albedo and membrane). A further study on this aspect is significant for development of effective and practicable residue definitions. Pomegranate fruit is characterized by a thick outer rind that encloses the soft and edible aril. Due to this typical nature of pomegranate fruits, the residue dynamics information generated for other fruits may not be applicable to it. The knowledge on fractionation of pesticide residues among different fruit parts of pomegranate is not available in literature, which could be the reason why the whole pomegranate fruits are considered for residue analysis in spite of the fact that the rind portion is non-edible or unsuitable for human consumption.

The studies on translocation of azoxystrobin, carbendazim and difenocazole showed that the residues of all the three fungicides were mostly accumulated in outer rind of the fruit. Small amount of azoxystrobin was found in the arils initially, which could occur as a result of the translaminar diffusion of this relatively polar chemical through rind (thickness ≈ 4.4 mm). The day-wise dataset of azoxystrobin residues in arils shows gradual decline, which could be attributed to the dissipation with passage of the time and concurrent dilution of the residues owing to juice filling and growth of the arils. In case of carbendazim and difenoconazole, the residues were not translocated into the albedo or any other parts of the fruit and the residues were confined to the rind of the fruit only. Thus, carbendazim and difenoconazole had very limited movement or penetration through the cuticle and their residues remained in the outer rind (thickness ≈3.06 mm) only.

For buprofezin and dimethoate, the residues were confined to outer rind only and were not further translocated into the internal parts of the fruit. However, imidacloprid which is systemic was found to translocate through plant tissues into the internal parts of the fruit. Food processing: a tool to pesticide residue dissipation: Processing of the fruits to juices and other processing products leads to large reductions in residue levels in the final produce. The dynamics of the residues during processing is an important aspect since it directly affects the acceptability of the produce. It is also relevant to explore alternative strategies (e.g., food processing) of saving the produce especially when pesticide residues are found in the final crop so that the produce does not incur major financial losses. It is therefore of significance to evaluate simple, cost effective strategies to ensure food safety of pomegranate fruits that are contaminated with pesticide residues. Food processing at domestic and industrial level could serve as suitable means to reduce the residue load on/in the final product and convert it to more cost-effective and acceptable forms. The processing of food commodities generally implies the transformation of the perishable raw commodity to value added product that has greater shelf life and is closer to being table ready.

The processing of the arils to juice does not have significant effect on the decontamination/degradation of the studied fungicides when analysed freshly. The residues of the fungicides were below their respective MRLs in the final product. Similar to the fungicides, the processing of the arils to juice does not have significant effect on the decontamination/degradation of the studied insecticides when analysed freshly. However, much of the residues were minimized during processing of arils to anardana. The residues of insecticides were minimized with the increase in treatment time. The residue levels came below the respective MRLs in the final product indicating processing of arils as one of the decontamination measures. Residues were below the respective EU-MRLs for fresh pomegranates and in the finished product.

Unit operations normally employed in processing food crops might reduce or remove residues of pesticides. Simple operations such as washing, peeling, blanching and cooking might play a significant role in the reduction of residues. Each operation has a cumulative effect on the reduction of the pesticides present. There is need to optimize the processing techniques with regard to pesticide residue dissipation and nutrient content. Substantial attention needs to be focused on addressing optimization of the processing techniques in a manner that leads to considerable pesticide residue dissipation but preserves most of the essential nutrients, thereby addressing the delicate balance between the two important parameters of food quality and safety. Thus, another important aspect of the study is based on the processed pomegranate commodities available in the market.

Methods of Residue Analysis

In modern day agriculture, agrochemicals belonging to different chemical classes are used for the control of diseases, pests, *etc.* In addition to the recommended chemicals, other chemicals might found source in the system via adjoining field drift, contaminated ground water, soil, and other agro-chemical inputs. Thus the regulations include setting up of MRLs for the wide varieties of these compounds viz. pesticides, growth regulators, herbicides, *etc.*

International organizations such as the EU or the Environmental Protection Agency [EPA] or domestic regulations (FSSAI) hence limit the amount of pesticides in food materials by setting MRLs with the aim of minimizing the consumer's intake of toxicants. The fulfillment of this legislation has led to development of MRLs at or below the lowest limit of analytical determination, usually ranging from 0.01-0.05 mg/kg (ppm). For effective implementation of monitoring programs there is also a requirement of fast, cheap and reliable method for multiresidue analysis with sufficient accuracy and reproducibility. Due to the low detection levels recommended by regulatory bodies and the complex nature of matrices such as pomegranates, efficient sample preparation and trace-level detection and identification are important aspects. Multiclass, multiresidue methods determination of pesticide residues in fruits (Savant et al, 2010; Banerjee et al, 2008) and processed products (Dasgupta et al, 2011) are indisputably one way of addressing the problem of pesticide determination. Sample preparation methods for faster turn-around time include extraction using organic solvents (acetonitrile, ethyl acetate, *etc.*) in presence of salts (magnesium sulphate or sodium sulphate) and are becoming popular in laboratories whereas gas chromatography mass spectrometry (GC- MS) and liquid chromatography mass spectrometry (LC-MS) are increasingly utilized for unambiguous identification and precise quantification of pesticide residues in compliance to the method validation guidelines of international bodies (AOAC, FG-SANCO of European Union, *etc.*).

Conclusion

Pomegranate is one of the important horticultural crops in India having high medicinal properties. The residues of pesticides in pomegranate are point of concern as far as exports and domestic food consumption are considered. The produce has to be residue free in order to fuilfil comprehensive and effective food legislations. This can be achieved by sincere adherence to the recommendations of the GAP and PHI. However immediate attention on the studies to generate PHIs and its implementation at the field level are of prime importance. This can definitely minimize the residue load of agrochemicals in pomegranate which in turn, will ensure safety to the health of the consumers and promote trade. The processing of pomegranate could be important in minimising the pesticide residue hazards. An effective implementation of residue monitoring programme (RMP) in pomegranate [initiated by Agricultural and Processed Products Export Development Authority (APEDA)] is essential to promote trade through establishment of food safety traceability system at farm/plot level. Utilization of weather forecasting based crop protection advisories can ensure need based pesticide applications, which in turn can minimize residue loads in the final produce and ensure food safety.

38

Entrepreneurship Development through Value Addition in Pomegranate: A Success Story

L.R. Tambade and A.S. Shelke,
KrishiVigyan Kendra, Solapur-413 255, Maharashtra

Introduction

Pomegranate fruits being rich source of minerals, vitamins and nutrients find wide application in traditional Asian medicines both in Ayurvedic and Unanic systems. Charak, the great medical physician of ancient India has prescribed a large number of formulations using almost every part of this plant in the treatment of dysentery, diarrhea, stomachache, inflammations, tapeworm, hynenole-tidosis, dyspepsia, bronchitis and cardiac disorders. These therapeutic properties are reported to be due to presence of betulic and uroslic acids and different alkaloids, *viz.*pseudopelletierine, pelletierine and some other basic compounds. A number of processed products can be manufactured by processing the fruits that are not marketable and get better prices. the processed products can be preserved for future use.

Pomegranate juice makes a delicious drink. On whole fruit basis the juice yield is about 42% while from grains the yield is about 70%. The juice can be extracted by using a spiral-type screw press without crushing the seeds. The juice is clarified by heating in a flash pasteurizer at 79-82^0 C cooling, settling for 24 hours racking up and filtering or decanting. The clear juice can be preserved by heat treatment or by using chemicals (600 ppm sodium benzoate).

The juice is highly nutritious and is recommended for patients suffering from gastric troubles. It contains 16.2% TSS and 0.35% acidity; total sugars, 12.93; reducing sugars, 12.65 and non-reducing sugars, 0.28% and 9.23 mg/100 g ascorbic acid.

Here we present a success story of a woman-Mrs. Vaishali Vishal Bele, whose husband is looking after farming, including pomegranate cultivation and she is involved in processing of unmarketable fruits.Here we present a brief description of the self employment of the lady farmer from Sangolawho has increased the annual income through processing of pomegranate.

Value Addition of Pomegranate: A Success Story

Pomegranate crop can be a boon and more so for small farmers for secure livelihood. This is exemplified by inspiring success story of a lady farmer from Sangola (Solapur) who ventured into pomegranate processing for self employmentand income generation. The detailed information is given below.

Name of the farmer: Mrs.Vaishali Vishal Bele
Age: 35 years
Education: H.S.C.
Address:A/P.: Kharatwadi, Tal. Sangola, Dist. Solapur
Cell. No.9764642132

01.	Family Land holding (ha.)	8 ha. land (Irrigated 4: Pomegranate, Ber, Drumstick,Fodder crops and Rainfed 4 ha.: Rabi Jawar, Wheat ,Gram)
02.	Size of the family	Big size (8 Members)
03.	Previous practice / Background	Working in own farm, hence earning was very negligible amount.
04.	Annual income before KVK,intervention (Rs)	30.000/- per year
05.	Name of unit & year of establishment	Arakta Fruit Juice CenterSangola Est. during 2010
06.	Source of fund for establishing the enterprise / unit	Self Investment of RS. 25000/ - during the year 2010. For purchase of Juicer, sealer and utensils
07.	Details of KVK intervention	
	a .Training	Participated in training on value added products from fruits & vegetables conducted by KVKduring sept.2010 Also Participated in training on FSSAI rules and Licensing procedure for Agri. Processing unit development at KVK Campus
	b. Demonstration	Participated in demonstration and training programme on preparation of pomegranate RTS & syrup & other fruit products

Contd...

	c. Technical services	Provided technical literature regarding preparation of fruits RTS packing and marketing of pomegranate.
	d. Follow up guidance	Helped for personality development, knowledge up gradation & linkages with marketing *etc.*
	e. Any other support	Developed l inkages with other state department, DRDA, MAV IM, Z.P. office, super market for marketing of products. Also facilitate her for getting FSSAI Licence
08.	Impact of intervention	
	a. Number of units established	01
	b. Size of unit	100 sq. ft.
c.	Production/unit/year	6000 lit.
	d. Annual Gross income (Rs)	3,40,000/-
	e. Annual Net Income/Unit (Rs)	1,10,000/-
f.	Economics of unit	Increase of income due to KVK intervention is up to 366%.
	g. No. of years enterprise is functional	4 years
	h. No. of full time employees (Other than entrepreneur)	01
	i. No. of part time/occasional employees	02
	j. Market detail/Sale of produce	Supplied produce to Super market, Government aided exhibition and Hotels, NGO, Beauty Parlour shop at Pune & lvfumbai.
09	Feedback	'I had gained the knowledge and confidence during the exposure visit organized by the KVK which was found veryuseful while searching the markets. I am really thankful to KVK, Solapur for developing me as a successful entrepreneur'
10	Social impact/Recognition	KVK Solapur recommend her as a HTA (Honorary Training Associate.) of KVK

39

Pomegranate: A High Potential Crop in Agro-Industry of Maharashtra based Entrepreneurship Development

Umakant Dangat

Commissioner, Agriculture & Managing Director, Maharashtra State Horticulture & Medicinal Plants Board, Pune-411 005, Maharashtra

Introduction

The area under fruit orchards in Maharashtra has considerably increasedeversince the fruit plantation scheme was clubbed with an employment guarantee scheme for the rural upliftment,. Subsequently ambitious launch of National Horticulture Mission by the Government of India for the holistic development of horticulture sector in the country and its quick implementation by the government in the state, made Maharashtra a leading state in the production of different types of fruits and other horticulture crops like vegetables, flowers and medicinal plants. Presently the state of Maharashtra is well ahead in area under cultivation of fruit crops like Mango, grapes, banana, guava, oranges, papaya, custard apple and pomegranate. Though the state has more area under various fruit crops compared to the other states, there seems a tremendous scope in increasing the productivity and tapping the potential in processing or export of the quality produce.

Among the major fruit crops in Maharashtra, pomegranate is gaining very much importance in recent times because of its high tolerance to the stressful conditions, high returning capacity per unit area as compared to the other fruit crops, more sturdiness of fruits in long distance transport and hence more export potential. The global trend of increased demand for pomegranate as fresh fruit or as derived products is growing at an impressive pace. It is an ideal crop for the sustainability of small holdings as it is well adapted to different topographies, soil and agro-climatic condition prevailing in arid and semi-arid regions of Maharashtra. Pomegranate (*Punica granatum* L.) is considered to have its native in the region from Iran to northern India with wild plants in many forests of these areas. It is a high value crop and its entire tree is of great economic importance. All parts of pomegranate tree have great therapeutic value and are variously being used in pharmaceutical, tanning and dying industry. Apart from its demand for fresh fruits and juice, the processed products like wine and candy are also getting importance in world trade. It has a deep association with the culture of Mediterranean region and Near East where it is savored as a delicacy and is an important dietary component and greatly appreciated for its medicinal properties. It has beneficial phytochemicals which have curative potential of many diseases like coronary heart disease, cancers of skin, breast, prostate and colon, inflammation, hyperlipidemia, diabetes, cardiac disorder, hypoxia, ischemia, aging, brain disorders and AIDS.

Present Scenario

Realizing its potential in generating high profits, there has been steady increase in area under pomegranate cultivation. India is a leading pomegranate producer with nearly 50% of world's production. The other major producers are Spain, USA, Iran, Turkey, Israel, *etc.* Total area under pomegranate in India was 1.13 lakh in 2012-13, but according to advanace estimates in 2013-14 it has increased to 1.26 lakh ha. In 2009-10, new plantations resulted in increase in area, however, in 2010-11 again the area dropped down. This could be due to removal of old bacterial blight affected orchards, as by now several farmers were confident of managing bacterial blight due to several demonstrations taken by NRCP, Solapur, MPKV, Rauhri and MAU, Parbhani in Maharashtra, under a network project on mitigating bacterial blight, funded by NHM, Ministry of Agriculture, GOI, New Delhi from 2008 to 2012. Thereafter, constant increase in pomegranate area was observed (Table 39.1), lower productivity in these years could be due to inclusion of area of new orchards which have not yet come to bearing. In order to mitigate bacterial blight of pomegranate a financial package was also sanctioned by the Ministry of Agriculture for 2008, 2009 and 2010 and hundred per cent pomegranate area (infected or free) was covered under the package. The package was sanctioned for the states of Maharashtra, Karnataka, Andhra Pradesh and Tamil Nadu.

Table 39.1: Pomegranate Area Production and Productivity in India 2005-06-2013-14

Year	Area x1000ha	Production x1000MT	Productivity t/ha
2005-06	111.6	809.2	7.25
2006-07	116.9	839.7	7.18
2007-08	123.6	884.1	7.15
2008-09	109.21	807.17	7.39
2009-10*	127.16	820.97	6.46
2010-11	107.3	743.1	6.90
2011-12	112.2	772.4	6.88
2012-13	113.25	744.96	6.58
2013-14 (Adv.estimate)	126.27	822.80	6.59

Source: Area and production data from Indian Horticultural Data Base 2009.-2014 www.nhb.gov.in

Although India is having largest area and production under pomegranate, its productivity is quite low (6.88t per ha) compared to other pomegranate growing countries like Spain (18.5t per ha), USA (18.3t per ha) and Turkey (11.3t per ha).

Table 39.2: Pomegranate Area, Production and Productivity in India (2014)

State	Area (x 1000 ha)	Production (x 1000 MT)	Productivity (t ha-1)
Maharashtra	90	477.0	5.3
Karnataka	15.5	154.8	9.98
Gujarat	7.4	79.0	10.4
Andhra Pradesh + Telangana	6.4	64.2	10.0
Madhya Pradesh	2.4	25.3	10.64
Himachal Pradesh	2.0	0.7	0.32
Rajasthan	1.1	4.1	3.86
Jharkhand	0.5	2.2	4.25
Tamil Nadu	0.4	13.1	32.7
Orissa	0.2	0.9	3.4
Jammu & Kashmir	0.1	0.3	3.07
Chhattisgarh	0.1	0.5	3.64
Nagaland	0.1	0.7	6.0
India (Total)	126.27	822.80	6.51

Source:- Indian Horticultural Data Base 2014, National Horticulture Board, Ministry of Agriculture, Government of India, Gurgaon (Haryana), www.nhb.gov.in

As per recent advance estimates for the year 2014 (Table 39.2), available at the National Horticulture Board of India website http://nhb.gov.in, total area under pomegranate in India is 1.27 lakh ha out of which 90,000 ha is in Maharashtra. States like Tamil Nadu, Rajasthan, Punjab, Haryana and Chhattisgarh are also venturing into pomegranate industry in order to diversify their existing cropping system.Considering the productivity, in Maharashtra it is 5.3t per ha. Highest productivity is in Tamil Nadu (32.7t/ha). The states like Karnataka, Gujarat, Andhara Pradesh, Telangana and Madhya Pradesh have restricted area with a reasonably good productivity of 10.0-10.64t/ha. Large scattered holdings, improper management and non- availability of varieties resistant to disease and insect pests are mainly responsible for low yields.

In Maharashtra Solapur, Nashik and Sangliare major pomegranate growing districts covering 75-80% area in Maharshtra. However, the farmers in other districts like Beed, Jalana, Aurangabad, Pune, Osmanabad, Satara, Ahemadnagar, Dhule, Jalgaon, Parbhani, Yeotmal, Akola, Buldana, Nanded, Amrawati *etc.* are showing their keen inclination towards this lucrative crop.

Export from Maharashtra

The state of Maharashtra is a leading state in exporting pomegranate fruits and its products along with other horticulture produce. Its pomegranate contribution which mainly consists of fresh fruits is more than 60% of the total export from all over India. Strategically an important port of India – Mumbai being a part of Maharashtra state, provides an advantage to the state growers in exporting the fresh produce to most of the gulf countries, Europe and America.

Table 39.3: Pomegranate Export from India from 2001-02 to 2014

	2004-05	2005-06	2006-07	2007-08	2008-09	2009-10	2010-11	2011-12	2012-13	2014
Export (x 1000 tonnes)	12.035	19.652	21.67	35.175	34.811	33.415	18.65	30,16	36,03	31.33
Value (x million rupees)	259	567	796	912	1146	1194	691	1473	2345	2985
Average export rate Rs. Per Kg	21.5	28.9	36.7	25.9	32.9	35.73	37.75	48.84	65.08	95.28

Source: Source: Agricultural & Processed Food Products Export Development Authority, Ministry of Commerce and Industry, www.apeda.gov.in

The major export destinations for Maharashtra's pomegranate are UAE, The Netherlands, UK, Belgium and Saudi Arabia. Over the period the export destinations for pomegranate more or less remained same except for the shares of Netherlands and Belgium which have become important destination for pomegranate export. The export trends during last 10 years (Table 39.3) show increasing trend in monetary value of pomegranate, hence, pomegranate cultivation has a bright future.

Demand in International Market

The highest prices for the fresh "pomegranate" are obtained in European and North American market. On examining the prices obtained by suppliers to the market of the West, where they are relatively high, it can be generalized that the preferred fruit characters more or less answer to the description of the American variety "Wonderful". Our soft seeded variety "Bhagwa" developed by MFKV – Rahuri is giving tough competition to "Wonderful" which has dark red exterior as well as interior-the arils. Its taste is sweet with a touch of acidity which is in contrast to 'Wonderful' having higher acidity and hard seeds or other varieties of Spain and Turkey, which are characterized by a sweet taste with no tartness.

In addition to all these, the demand for this fruit also originates from the industries that are making use of pomegranate fruit essence in high demand products for enriching them with anti- oxidants. These include the beverage, food, cosmetics food additives and the counter medicine industries.

Strengths and Concerns

Strengths

Pomegranate is a fruit crop of arid and semiarid tropics. It requires hot, dry temperature and prefers light to medium type of soil. These conditions are very common in a major part of Maharashtra and hence being a high return giving crop, pomegranate is gradually taking place of other established horticulture and fruit crops.

Being a fruit of great nutritional and medicinal value as compared to the other fruit crops, pomegranate has increasing demand in domestic as well as international market.

Fruits of pomegranate being tough with good shelf life after harvest as compared to the other fruits, can sustain abrasions in long distance transport and have a good export potential to the distant countries.

Being preferred in the market in various forms, pomegranate has better scope of value addition and processing.

Major work on development of new varieties has been done in Maharashtra. The varieties available in India or Maharashtra produce very attractive red colour fruits. Even the aril colour is also very attractive. The seeds are soft and arils taste sweet. This makes the Indian produce more preferable in the international market than the produce from other countries like Iran or Spain.

Bhagwa variety has high acceptance in the European market.

This crop can becultivated on marginal land with proper nutritional and irrigation management to control flowering and fruiting. This enables Maharashtra to produce and supply fruits throughout the year. This is a very plus side of the crop from the marketing point of view as compared to the other established fruit crops.

Considering the huge potential in this fruit crop, the government of Maharashtra has established an agriculture export zone for enhancing exports of pomegranate.

There is avstrong research support for scientific cultivation of pomegranate from

Institutes like National Research Center onPomegranate, Solapur, MPKV, Rahuri in Maharashtra and IIHR, Bangalore in Karnataka state.

Pomegranate co-operative societies from Maharashtra state have formed an apex cooperative namely MAHAANAR.

Pomegranate export facility center has been set up in Baramati area with modern handling facilities.

Farmers have been trained for export quality production and have registered with GLOBALGAP certification.

MSAMB has recently obtained brand name i.e. "MAHAPOM".

Various private traders are also encouraging farmers to go for this crop and export of the produce.

Demand in the international market and comperatively high returns per unit area, has widened the scope for extension of area and enhancement of pomegranate production in Maharashtra. However, increasing the productivity and consistent quality production is great challenge in front of scientists working in this field and the farmers. In order to stand in the domestic as well as international market, sincere efforts and scientific as well as commercial approach is required to retain our number one position among other states in respect of pomegranate crop.

Concerns

Following are the concerns which need to be addressed by all related to pomegranate

crop management and marketing.

Because oflarger area under pomegranate, efficient adoption of good management practices many a times become a difficult proposition which may be a hindrance for the growth of pomegranate productivity.

Heavy or saline, lands are being put under pomegranate cultivation, particularly in Vidharbha and some other parts of Maharashtra. Such soils are highly unsuitable for this crop. Hence, either proper extension service should be provided in these areas or research should be done for managing this crop in such type of soil conditions.

It is grown as intensive crop on a commercial basis, heavy chemical and fertilizer inputs gets poured in the pomegranate orchards. This gives good production in terms of quantity but at the same time the final produce carries the residual traces of the chemicals applied. This hinders the opportunity of exporting as the product needs to go stringent quality tests before entering in foreign countries.

Widespread occurrence of some diseases like bacterial blight and wilt which were of little concern to the pomegranate growers in the past, in recent times have become a cause of concern.

An availability of disease free planting material, varieties having resistance to important diseases and pests is a matter of serious concern. Even the recent availability of tissue cultured pomegranate plants through some private laboratories doesn't provide guarantee about their disease free nature. Here, a thorough scientific study and appropriate regulations for quality maintenance in nursery practices are required to provide the best planting material in order to avoid further consequences.

Conserving native pomegranate biodiversity and explorations of new, non-traditional potential areas for cultivation are some of the issues requiring immediate attention.

Creation of adequate marketing, storage and export facilities also need intervention of the policy makers.

Measures for Enhancing Competitiveness for Export

Following measures if adopted are can enhance competitiveness of Pomegranate produce in overseas markets:

Competition of India with regard to export of pomegranates is with Spain and Iran, which are nearer to European countries who import maximum quantity. Our efforts need to be focused on reducing the cost of production with an increasing productivity.

Modern pack house facilities need to be established on large scale.

We have already embarked upon building up quality and branding its product in order to compete with Spain and Iran. The brand needs to be popularized aggressively.

Pomegranate supplies from Spain and Iran to Europe taper from January onwards and therefore, supplies from India need to be concentrated during February to July months with the help of *Hast* and *Ambe bahar* when there will be no competition from Spain.

Efforts need to be made to popularize this fruit in Canada, U.S.A., South American countries *etc.* by holding fruit fairs, exhibitions etc, as there is good price realization also from these countries. Similarly, efforts need to be accelerated in popularizing pomegranates in Australia, Korea, Japan, *etc.*

In order to achieve maximum productivity and returns the postharvest management and value addition should be properly studied.

Futuristic Approach

The experiences of many countries world over suggest that export orientation of horticultural produce is one of the prerequisite for its contribution to global trade. Further it has also been observed that export orientation of horticultural produce is sustained when complemented with a sizable processing industry and strong internal market. As we lack these prerequisites, much success on the export front has not been achieved in spite of the fact that Maharashtra is the largest producer of pomegranate not only in India by globally. To give an impetus to pomegranate export it is essential to follow five-pronged strategy consisting of product segmentation, market diversification, market penetration, value addition and infrastructure up-gradation. Moreover, the growers need to be given risk protection against market as well as production. Two important requirements for successful development of pomegranate sector are to ensure consistency in supply and provide recorded and demonstrated traceability of products. Thus, production strategies play central role in the all-round development of this sector. Now it is crucial that innovation in both production as well as protection should be focused on developing sustainable production techniques for pomegranate and also on value addition to the product through packaging and processing.

Further, the development of pomegranate industry/sector needs to be accompanied by the growth of food processing industry in the nearby area. The opportunity also exists to promote this industry by intensifying production of suitable variety for products like juice and concentrates *etc.* Thus, the production strategy should be aimed at meeting not only domestic and export demand but also of processed products. This sector requires to be developed as an organized industry and will have to be run collectively by all the stakeholders with grower as the entrepreneurs.

More over, the marketing cost is almost 50% of the total production cost, thus there is a need to set up institutional agencies that can lend advance credit to the growers and motivate them to market the produce themselves. There is also an immediate need to integrate the production, marketing and processing through involvement of institutional structure which will co-ordinate between demand and supply of inputs as well as produce to have maximum benefit from pomegranate cultivation. Considering above aspects institutional setup in PPP-IAD mode can be viable & efficient alternative to achieve the above objectives of development of pomegranate industries.

Government Schemes to Foster the Development of Pomegranate Industries

Considering the necessity of developing horticulture industry the following schemes need tobe implemented by state/central government:

a) Mission for integrated development of horticulture (MIDH):

b) The components like establishment of new gardens, trainings, exposure visits of growers for capacity building and mechanization, pack house, cold-storage, processing, marketing infrastructure *etc.* can be integrated to establish, viable institutional arrangement for value chain development.

c) Farmers Producer Organization/Companies can be establish involving 1000-5000 farmers for which assistance can be sought through RKVY for various components requirement for establishment of value chain.

d) Ministry of Food Processing Industry (MOFPI

e) MOFPI runs various schemes necessary for food processing and establishment of integrated value chain.

g) In addition to above, assistance can be sought through schemes like national mission for micro irrigation, EGS horticulture, MACP, SFAC, NABARD *etc.*

Focus Areas

I. Research and Development (R&D) : In addition to regular R&D work being carrying out at various research stations the special focus needs to be given on following points -

a) Bacterial Blight and wilt like important diseases needs to be addressed.

b) Varietal improvement to cater tothe needs of domestic as well as export market and processing.

c) Improvement of yield and quality of fruits so as to stand in international market.

d) Tissue Culture and other suitable means of propagation.

e) *Bahar*treatments need to be studied more effectively so as to get assured flowering.

To conclude I wo uld say the journey of pomegranate industry has just begun, there is a long way to go.

References

Ahuja, D. B., Sharma, J., Suroshe, S. S. 2013. CROPSAP (Horticulture) team of e' pest surveillance: 2013: Pests of Fruits (Banana, Mango and Pomegranate) - 'e' Pest Surveillance and Pest Management Advisory, (Ed. DB Ahuja) Published by National Centre for Integrated Pest management, New Delhi and Department of Horticulture, Commissionerate of Agriculture (Horticulture), MS, Pune. 67 pp.

Aly Khan, Shaukat, S. S., Mian Sayed. 2011. Management of plant nematodes associated with pomegranate (Punica granatum L.) using oil-cakes in Balochistan, Pakistan. *Indian Journal of Nematology*, **41**(1): 1-3.

Ananthakrishnan, T. N. 1993. Bionomics of thrips. *Annu. Rev. Entomol.* **38**, 71–92.

Anonymous. 2009. Crop Management. DARE/ICAR Annual Report, 2008–09. p 54

Anonymous. 2012. NRCP Annual Report, 2011-12, National Research Centre on Pomegranate, Solapur-413 255, Maharashtra. p 1.

Anonymous. 2013a. NRCP Annual Report, 2012-13, National Research Centre on Pomegranate, Solapur-413 255, Maharashtra. p 83.

Anonymous. 2013b. NRCP Annual Report, 2012-13, National Research Centre on Pomegranate, Solapur-413 255, Maharashtra. p 82.

Austin, G. D. 1956. Historical review of shot-hole borer investigations. *Tea Quarterly*. **27**: 97–102.

Bagle, B. G. 2011. Studies on varietal reaction, extent of damage and management of anar butterfly, *Deudorix (=Virachola) isocrates* Fab., in pomegranate. *Acta Horticulturae*, **890**: 557-559.

Bagle, B.G. 1993. Seasonal incidence and control of thrips *Scirtothrips dorsalis* Hood in pomegranate. *Indian J. Entom.* 55:148-153.

Balikai, R. A., Kotikal, Y. K., Prasanna P. M. 2011. Status of Pomegranate pests and their management strategies in India. *In*: Proc. IInd IS on Pomegranate and Minor, including Mediterranean Fruits (Eds.: M.K. Sheikh et al.). *Acta Hort.* **890:** 569-584.

Balikai, R. A., Kotikal, Y.K., Prasanna P. M. 2009. Status of Pomegranate pests and their management strategies in India. ISHS Acta Horticulturae 890: II International Symposium on Pomegranate and Minor - including Mediterranean - Fruits: ISPMMF

Baptist, B. A. 1944. The fruit-piercing moth (*Othreis fullonica* L.) with special reference to its economic importance. *Indian Journal of Entomology*, **6**: 1-13.

Bayhan, E., Olmez-Bayhan, S., Ululsoy, M. R., Brown, J. K. 2005. Effect of temperature on the biology of *Aphis punicae* (Passerini) (Homoptera: Aphididae) on pomegranate. *Environmental Entomology*, **34**(1): 22-26.

Beaver, R. A. 1989. Insect-fungus relationships in the bark and ambrosia beetles. In: N. Wilding, N.M. Collins, P.M. Hammond, & J.F. Webber (Eds), Insect-fungus Interactions. Academic Press, New York, NY, USA, pp. 121-144.

Bellows, T. S., Paine, T. D., Arakawa, K. Y., Meisenbacher, C., Leddy, P. and Kabashimo J. 1990. Biological control sought for ash whitefly. *California Agriculture*, **44**: 4-6.

Bhatnagar, S. P. 1957. Description of new and records of known Chalcidoidea (Parasitic Hymenoptera) from India. *Indian Journal of Agriultural Science*, **21**:155-178.

Bhatt J, Chaurasia RK, Sengupta, SK. 2002. Management of *Meloidogyne incognita* by *Paecilomyces lilacinus* and influence of different inoculums levels of *Rotylenchus reniformis* on betelvine. *Indian Phytopath*, **55**(3): 348-350.

Bhumannavar, B. S. 2000. Studies on fruit piercing moths (Lepidoptera: Noctuidae) – species composition, biology and natural enemies. Ph. D. Thesis submitted to University of Agricultural Sciences, GKVK, Bangalore, 181 pp.

Bhumannavar, B. S. and Viraktamath, C. A. 2000. Biology and behaviour of *Euplectrus maternus* Bhatnagar (Hymenopera: Eulophidae), an ectoparasitoid of *Othreis* spp. (Lepidoptera: Noctuidae) from southern India. *Pest Mngt. in Hort. Ecosyst.,* **6**(1): 1-14.

Bhumannavar, B. S. and Viraktamath, C. A. 2001a. Larval host specificity, biology, adult feeding and oviposition preference of the fruit piercing moth, *Othreis homaena* Hubner (Lepidotera: Noctuidae) on different Menispermaceae host plants. *Journal of Entomological Research*, **25**(3): 217-233.

Bhumannavar, B. S. and Viraktamath, C. A. 2001b. Seasonal incidence and extent of parasitisation of fruit piercing moths of the genus Othreis (Lepidoptera: Noctuidae). *Journal of Biological Control*, **15**(1): 31-38.

Bhumannavar, B. S. and Viraktamath, C. A. 2012. Biology, ecology and management of fruit piercing moths (Lepidoptera: Noctuidae). *Pest Management in Horticultural Ecosystems*, **18**(1): 1-18.

Bindra, O. S. 1969. Insect pests of citrus, review of work done, lacunae in our knowledge and future line of work. ICAR Workshop Fr. Res., PAU, Ludiana, April 1969, pp 128-30.

Biradar, A. P., Jagginavar, S. B. and Sunitha, N. D. 2005. Management of stem borer, *Coelosterna scrabrator* Fabr. (Coleoptera: Cerambycidae) in pomegranate. *International Journal of Agricultural Sciences*, **1**(1):16-17.

Bobat, A. 1997. Pheromone trials for biochemical control of the striped ambrosia beetle, *Trypodendron lineatum* (Olivier). - *Trop. J. Agric. For.* **21**(6): 599-603.

Carneiro, R. M. D. G., Almeida, M. R. A., Cofcewicz, E. T., Magunacelaya, J. C., Aballay, E. 2007. Meloidogyne ethiopica, a major root-knot nematode parasitizing Vitis vinifera and other crops in Chile. Nematology, **9**(5): 635-641.

Chandra, R., Suroshe, S. S, Sharma, J., Marathe, R. A., and Meshram, D. T. 2011. *Pomegranate growing manual.* NRC on Pomegranate, Solapur, Maharashtra. 57 pp

Chandrasekaran, J., Azhakiamanavalan, R. S. and Louis, I. H. 1990. Control of grapevine borer, *Coelosterna scabrator* F. (Cerambycidae: Coleoptera). *South Indian Horticulture*, **38**(2):108.

Chinniah, C. and Mohansundaram, M. 1995. Record of new ascid mites (Ascidae:Acari) infesting insects in Tamil Nadu, India. *Entomon*, **20**(3 & 4): 233-236.

Danthanarayana, W. 1968. The distribution and host-range of the shot-hole borer (*Xyleborus fornicatus*) of tea. *Tea Bulletin*, 61-69.

Danthanarayana, W. 1973. Host plant-pest relationships of the shot-hole borer of tea (*Xyleborus fornicatus*) (Coleoptera: Scolytidae). *Entomologia Experimental & Applicatta.* **16**: 305-312.

Dhooria, M. S. and Sandhu, G. S. 1973. Occurence of *Tenuipalpus punicae* Pritchard and Baker (Tenuipalpidae: Acarina) on Pomegranate in India. *Current Science*, **42**(5): 179-180.

Dodia, J. F., Yadav, D. N. and Patel, R. C. 1986. Management of fruit sucking moth, Othreis fullonica Cl. (Lep.: Noctuidae) in citrus orchards at Anand (Gujarat). Gujarat AgriculturalUniversity Research Journal, **11**(2): 72-73.

Eisenback, J. D. & Triantaphyllou, H. H. 1991. Root-knot Nematodes: Meloidogyne species and races. In: Manual of Agricultural Nematology, (Ed: W. R. Nickle). Marcel Dekker, New York. pp. 281 – 286.

Elmore, C. L., Stapleton, J. J., Bell, C. E., and DeVay, J. E. 1997. Soil Solarization: A Non-pesticidal Method for Controlling Diseases, Nematodes and Weeds. Oakland: *Univ. Calif. Agric. Nat. Res. Publ.* **21377**: 13 p.

Fand, B. B., Gautam, R. D. and Suroshe, S.S. 2010a. Comparative biology of four coccinellid predators of solenopsis mealybug, *Phenacoccus solenopsis* Tinsley (Hemiptera: Pseudococcidae). *Journal of Biological Control,* 24(1): 35-41.

Fand, B. B., Gautam, R. D. and Suroshe, S.S. 2010b. Effect of developmental stage and density of *Phenacoccus solenopsis* Tinsley (Hemiptera: Pseudococcidae) on the predatory performance of four coccinellid predators. *Journal of Biological Control,* **24** (2): 110-115.

Fand, B. B., Gautam, R. D. and Suroshe, S.S. 2011. Suitability of various stages of mealybug, Phenacoccus solenopsis (Homoptera: Pseudococcidae) for development and survival of the solitary endoparasitoid, Aenasius bambawalei (Hymenoptera: Encyrtidae). *Biocontrol Science and Technology,* **21**(1): 51-55.

Fasih, M. and Srivastava, R. P. 1988. Natural occurrence of *Beauveria bassiana* an entomogenous fungus on bark eating caterpillar, *Indarbela* spp. *Indian Journal of Plant Pathology,* **6**(1): 11-16.

Fay, H. A. C. and K. H. Halfpapp. 1993. Non-odorous characteristics of lychee (*Litchi chinensis*) and carambola (*Averrhoa carambola*) pertaining to fruit-piercing moth susceptibility. *Australian Journal of Experimental Agriculture,* **33**: 227-231.

George Mathew. 1997. Management of the bark caterpillar *Indarbela quadrinotata* in forest Plantations of *Paraserianthes falcataria. KFRI Research Report,* **122**. 24 pp.

Gillespie, P. S. 2000. A new whitefly for NSW - The ash whitefly. NSW Agriculture. http://www.agric.nsw.gov.au/Hort/ascu/insects/ashwf.htm (21 September 2010).

Gorantiwar S D, Meshram D T and Mittal H K. 2011. Water requirement of pomegranate for Ahmednagar district part of Maharashtra. ***Journal of Agrometeorology.*** 13(2):123-127.

Haneef M, Kaushik R A, Sarolia D K, Mordia A and Dhakar M. 2014. Irrigation scheduling and fertigation in Pomegranate cv. Bhagwa under high density planting system. ***Indian Journal of Horticulture.*** 71(1):45-48.

Hartman, K. M. and J. N. Sasser. 1985. Identification of Meloidogye species on the basis of differential host test and perineal-pattern morphology. pp. 69-76 in K. R. Barker, C. C. Carter, and J. N. Sasser, eds. An advanced treatise on Meloidogyne, vol. 2. Methodology. North Carolina State University Graphics, Raleigh.

Hill, D. S. 2008: Pests of crops in warmer climates and their control. Springer, 704 pp.

Holland, D., Hatib, K., and Bar-Ya'akov, I. 2009. Pomegranate: botany, horticulture, breeding: Janick, J. (Ed.), John Wiley and Sons, New Jersey, *Horticultural Reviews,* **35**: 127-191.

Hooks, C. R. R., Wang, K. H., Ploeg, A. and McSorley, R. 2010. Using marigold (*Tagetes* spp.) as a cover crop to protect crops from plant-parasitic nematodes. *Applied Soil Ecology*, **46**: 307-320.

Intrigliolo D S, Nicols E, Bonet L, Ferrer P, Alarcon J J and Bartul J. 2011. Water relations of field grown pomegranate trees (*Punica granatum* L.) under different irrigation regimes. *Agricultural Water Management.* **98**:691-696.

Jagginavar, S. B. Sunitha, N. D. Paitl, D. R. 2008. Management strategies for grape stem borer Celosterna scrabrator Fabr. (Coleoptera: Cerambycidae). *Indian Journal of Agricultural Research*, **42**(4): 307-309.

Jagginavar, S. B., Naik, L. K. 2001. Seasonal activity of the pomegranate shot-hole borer, *Xyleborus perforans* (Wollastan) (Coleoptera: Scolytidae). *Pest Management in Horticultural Ecosystems*, **7**(2): 141-146.

Jain, P. C. and Bhargava, M. C. 2007. Entomology: Novel Approaches. New India Publishing, 533 p.

James, S. P., Babu, A., Selvasundaram, R., Muraleedharan, N. 2007. Field evaluation of traps for attracting shot hole borer. *Newsletter - UPASI Tea Research Foundation*; **17**(1): 3.

Jayanthi, P. D. K, Abraham Verghese and Nagaraju, D. K. 2009. Studies on feeding preference of adult fruit sucking moth, Eudocima (Othreis) materna (L.): A clue for devising trap cropping strategies. *Pest Management in Horticultural Ecosystems*, **15**(2): 107-113.

Jayanthi, P. D. K. and Verghese, A. 2010. Natural parasitization of larvae of fruit piercing moth, *Eudocima* (*Othreis*) *materna. Insect Environment*, **16**(2): 67.

Joshi, B. D., Tripathi, C. P. M. and Joshi, P. C. 2009. Biodiversity and environmental management. APH Publishing, 176 p.

Kambrekar, D. N. And Kalaghatagi, S. B. 2012. Management of Anar butterfly in pomegranate. The Hindu, 23rd August.

Kar G and Kumar A. 2007. Effects of irrigation and straw mulch on water use and tuber yield of potato in eastern India. *Journal of Agricultural Water Mangement.* **94**(109):116-120.

Khan, M. R. and Goswami, B.K. 2002. Evaluation of *Paecilomyces lilacinus* isolate 2 against Meloidogyne incognita infecting tomato. *Int. J. Nematology*, **12**:111-114.

Khan, M. R., Jain, R. K., Ghule, T. M., and Pal, S. 2014. Root knot Nematodes in India - a comprehensive monograph. All India Coordinated Research Project on Plant Parasitic nematodes with Integrated approach for their Control, Indian Agricultural Research Institute, New Delhi. pp. 78 + 29 plates.

Kiewnick, S. and Sikora, R. A. 2006. Biological control of the root-knot nematode Meloidogyne incognita by *Paecilomyces lilacinus* strain 251. *Biol. Control.* **38**:179-187.

Kinawy, M. M., Al-Waili, H. M. and Almandhari, A. M. 2008. Review of successful biological control programmes in Sultanate of Oman. Egyptian Journal of Biological Pest Control, **18**(1): 1-10.

Krishna Rao, V. and Krishnappa, K. 1995. Integrated management of *Meloidogyne incognita-Fusarium oxysporum* f.sp. ciceri wilt disease complex in chickpea. *Int. J. Pest Management*, **41**: 234-237.

Kumar, K. and Lal, S. N. 1983. Studies on the Biology, Seasonal Abundance and Host-Parasite Relationship of Fruit Sucking Moth *Othreis fullonia* (Clerk) in Fiji. *Fiji Agric. J.*, **45**(2): 71-77.

Lal, K. B. 1953. Annual report on the scheme for investigations on some serious insect-pests attacking fruits and fruit trees grown in the plains of Uttar Pradesh, 1951-52.

Ma YanFen, Yuan ShengYong, Xiao Chun, Hu XianQi. 2013. Screening of trap plants of pathogenic root-knot nematode in pomegranate orchard [Chinese]. *Journal of Henan Agricultural Sciences*, **42**(4): 99-102.

Mani, M and Krishnamoorthy, A. 1995. Natural enemies of *Siphoninus phillyreae* (Homoptera: Aleurodidae) and *Aphis punicae* (Homoptera: Aphididae) on pomegranate. *Entomon*, **20**(1): 31-34.

Mani, M. and Krishnamoorthy, A. 2000. Biological suppression of mealybugs *Planococcus citri* (Risso) and *Planococcus lilacinus* (Ckll.) on pomegranate in India. *Indian Journal of Plant Protection*, **28**(2): 187-189.

Masarrat Haseeb and Shashi Sharma. 2007. Studies on Incidence and Crop Losses by Fruit Borer Deudorix isocrates (Lep: Lycaenidae) on Guava. *In*: Proc. I[st] IS on Guava (Eds. G. Singh et al.). *Acta Hort* (ISHS), **735.**

Methew, G. and Rugmini, P. 1998. Control of bark eating caterpillar in forest plantations of *Paraserienthes falcataria*. *Indian Journal of Environment and Toxicology*, **8**(1): 37-40.

Mohite, A. S., Umarkar, S. P. and Shinde, J. S. 1995. Laboratory screening of some organosynthetic insecticides against third instar larvae of fruit-sucking moth, *Othreis materna* (L.) (Lepidoptera: Noctuidae). *Indian Journal of Entomology*, **57**(2): 89-93.

Mote, U. N. Tambe, A. B. 1990. Insecticidal control of shot-hole borer, *Xyleborus fornicatus* Eichhoff on pomegranate. *Plant Protection Bulletin* (Faridabad), **42**(1-2):9-12.

Mote, U. N. Tambe, A. B. 2000. Effective and economic management of shot-hole borer on pomegranate. *Journal of Maharashtra Agricultural Universities*, **25**(2):155-157.

Mote, U. N., Tambe, A. B. 1990. Chemical control of bark eating caterpillar *Indarbela quadrinotata* (Walker) in pomegranate. *Plant Protection Bulletin* (Faridabad)m, **42**(3-4): 7-8.

Mote, U. N., Tambe, A. B. and Patil, C. S. 1991. Observation on incidence and extent of damage of fruit sucking moths on pomegranate fruits. *Journal of Maharashtra Agricultural University,* **16**(3): 438-439.

Motsinger, R. E., Moody, E. H. and Gay, C. M. 1977. Reaction of certain French marigold (Tagetes patula) cultivars to three Meloidogyne spp. *Journal of Nematology,* **9**: 278.

Muniappan, R., Denton, G. R. W., Marutani, M., Lali, T. S. and Kimmons, C. A. 1993. Fruit piercing moths in Micronesia and their natural enemies. *Micronesica Suppl.,* **4**: 33-39.

Murugan, M., Thirumurugan, A. 2001. Studies on the eco-friendly approaches to manage pomegranate fruit borer, *Deudorix isocrates* (Fab.). *Plant Protection Bulletin* (Faridabad), **53**(3/4): 13-15.

Nagaraja, H. and Ankita Gupta. 2007. A new species of *Trichogramma* (Hymenoptera: Trichogrammatidae) parasitic on eggs of pomegranate fruit borer, *Deudorix isocrates* (Fabricius) (Lepidoptera: Lycaenidae). *Journal of Biological Control,* **21**(2): 291-293.

Naik, L. K. Jagginavar, S. B. Biradar, A. P. 2011. Beetle enemies of pomegranate and their management. *Acta Horticulturae,* **890**: 565-568.

Nair, M. R. G. K. 1975. Insect and mites of crops in India. ICAR, New Delhi, 408 pp.

Narendran, T. C. and Sureshan, P. M. 1988. A contribution to our knowledge of Torymidae of India (Hymenoptera: Chalcidoidea). *Bollettino-del-Laboratorio-di-Entomologia-Agraria-Filippo- Silvestri,* **45**: 37-47.

Neeraj Kotwal, Kiran Kour and Kamaldeep Singh. 2012. Evaluation of some insecticides against pomegranate butterfly, *Deudorix isocrates* Fabricius. *Journal of Insect Science* (Ludhiana), **25**(2):208-209.

Paine, T. D., Raffa, K. F., and Harrington, T. C. 1997. Interactions among scolytid bark beetles, their associated fungi, and live host conifers. *Annual Reviews of Entomology,* **42**, 179-206.

Patel, A. N. and Patel, R. K. 2008. Biology of bark eating caterpillar, *Indarbela quadrinotata* (Walker). *Current Biotica,* **2**(2): 234-239.

Pawar, S. A., Mhase, N. L., Kadam, D. B. and Chandele, A. G. 2013. Management of Root-Knot Nematode, Meloidogyne incognita, Race-II infesting Pomegranate by using Bioinoculants. *Indian Journal of Nematology,* **43**(1): 92-94.

Perry, E. J. and Ploeg, A. T. 2010. Pest notes. *University of California, Publication,* **7489**: 1-5 pp.

Ramsingh, Jaswant Singh, J., Singh, R. and Singh, J. 1982. New record of *Aspergillus candidus* Link - a potential entomogenous fungus on Indarbela spp. *Science and Culture,* **48**(8): 282-283.

Roi Levin. 2005. Reproduction and identification of root-knot nematodes on perennial ornamental plants in Florida. M. Sc. Thesis submitted to the University of Florida, 197 pp.

Sandhu, G. S., Batra, R. C. and Sohi, A. S. 1980. Note on the incidence of fruit-sucking moths in the Punjab. *Indian Journal of Entomology*, **42**: 531-532.

Sasser, J. N. and Carter, C. C. 1985. Overview of the International *Meloidogyne* Project 1975-1984. In: An Advanced Treatise on *Meloidogyne*. North Carolina State University Graphics, 19-24.

Senguttuvam, T. 2000. Bark borer – A polyphagous pest on Agroforestry trees. *Insect Environment*, **6**(1): 28.

Sharma, D.D. and Kumar, H. 1986. How to control bark eating caterpillars'? *Indian Horticulture*, **31**(1): 25.

Sharma, J., Suroshe, S. S. and Shinde, Y. 2012. Diagnosis and integrated management of diseases and insect pests of pomegranate (In Marathi). *Extention Bulletin No. 7*, 58 pp

Shevale, B. S. 1991. Control of bark eating caterpillar, *Indarbela quadrinotata* (Walker) in pomegranate through insecticidal sprays. *Plant Protection Bulletin* (Faridabad), **43**(3-4): 31-32.

Shevale, B. S. 1994. Studies on control of pomegranate butterfly, *Virachola isocrates* Fabricius. *Plant Protection Bulletin* (Faridabad), **46**(1):19-21.

Shevale, B. S., and S.N. Kaulgud. 1998. Population dynamics of pests of pomegranate *Punica granatum* Linnaeus. Advances in IPM for horticultural crops. Proceedings 1st National Symp., Pest Management in Horticultural Crops: Environmental Implications and Thrusts, Bangalore, India, 15–17 Oct. 1997, 47–51.

Shevale, B. S., Khaire, V. M. 1999. Seasonal abundance of pomegranate butterfly, *Deudorix isocrates* Fabricius. *Entomon*, **24**(1): 27-31.

Singh, D. and D. C. Gupta. 1993. Evaluation of marigold cultivars/hybrids for resistance against *Meloidogyne javanica*. *Haryana Agricultural University Journal of Research*, **23**:156-159.

Singh, K. P., Bandopadhyay, P., Vaish, S. S., Makeshkumar, T. and Gupta, R. C. 2002. Growth and population dynamics of *Catenaria anguilluiae* in relation to oilcakes. *Indian Phytopath*, **55** (3): 286-289.

Singh, S. B. and Singh, H. M. 2000. Bioefficacy and economics of different pesticides against anar butterfly, *Deudorix isocrates* (Fabricious) (Lycaenidae: Lepidoptera) infesting aonla. *Indian Journal of Plant Protection*, **28**(2): 173-175.

Singh, S.P, Rao, N. S., Ramani, S. and Poorani, J. 2001. Research Highlights 200-2001. Bangalore, India, 21 pp.

Sivapalan, P. 1976. Control of shot-hole borer of Tea (*Xyleborus fornicatus* Eichh.). Final Report on Research conducted under grant Authorized by U.S. Public Law 480. Tea Research Institute, Talawakelle, Sri Lanka.

Somasekhara, Y. M., Ravichandra, N. G. and Jain, R. K. 2012. Bio-management of root knot (*Meloidogyne incognita*) infecting pomegranate (*Punica granatum* L.) with combination of organic amendments. *Research on Crops*, **13**(2): 647-651.

Speyer, E. R. 1917. Shot-hole borer of tea. *Tropical Agriculturist*, **49**: 17–21.

Subodhi Karunaratne, W., Vijaya Kumar, Jan Pettersson and Savitri Kumar, N. 2008. Response of the shot-hole borer of tea, *Xyleborus fornicatus* (Coleoptera: Scolytidae) to conspecifics and plant semiochemicals. *Acta Agriculturae Scandinavica, Section B - Soil & Plant Science*, **58** (4): 345-351.

Sunita. 2012. Intensity of anar butterfly, *Virachola isocrates* (Fabr.) with period and crop means. *Recent Research in Science and Technology*, **4**(9): 14-15.

Suroshe, S. S., Gautam, R. D., & Fand, B. B. 2013. Natural enemy complex associated with the mealybug, *Phenacoccus solenopsis* Tinsley (Hemiptera: Pseudococcidae) infesting different host plants in India. *Journal of Biological Control*, **27**(3): 204-210.

Suroshe, S. S., Singh, N.V., Maity, A., Meshram, D. T., Shinde, Y. R. and Pal, R. K. 2013. Fruit sucking moth and their management. 4p.

Susainathan, P. 1924a. Fruit-sucking moths of South India. Proceedings of 5th Entomological Meeting, Pusa, pp. 23-27.

Susainathan, P. 1924b. The fruit moth problem in the Northern Circars. *Agricultural Journal of India*, **19**: 402-404.

Swart, P. L., De Kock, A. E. and Rust, D. J. 1975. Fruit piercing moths. *Deciduous Fruit Grower*, **25**(4): 97-102.

Varma, A. N. and Khurana, A. D. 1977. Survey on the incidence of bark eating caterpillar (Indarbela sp.) on different fruit trees in Haryana. *Haryana Agricultural University Journal of Research*, **6**(2): 93-104.

Vasantharaj David, B. and Ananthakrishnan, T. N. 2004. General and Applied Entomology. Tata McGraw-Hill Education, 1183 pp.

Walgama, R. S. and Pallemulla, R. M. D. T. 2005. The Distribution of Shot-hole borer *Xyleborus fornicatus* Eichh. (Coleoptera: Scolytidae) Across Tea Growing Areas in Sri Lanka – A Reassessment. *Sri Lanka Journal of Tea Science*, **70**: 105-120.

Walgama, R. S., Senanayake, P. D., Abeysekera, A. R., Premathunge, A. K., Perea, R., Vitana, B. S., Sureshkumar, B. and Seram, C. De. 2008. Chemical control of shot-hole borer, *Xyleborus fornicatus* eichh. (coleoptera: scolytidae) in tea. *Tropical Agricultural Research & Extension*, **11**: 65-68.

Waterhouse, D. F. and K. R. Norris. 1987. *Othreis fullonia* (Clerk). pp. 240-249. In: Biological Control Pacific Prospects. Inkata Press; Melbourne. 454 pp.

www.extento.hawaii.edu/kbase/crop/type/othreis.htm, accessed on 02-08-02014.

www.kiengiangbiosphererereserve.com.vn/project/uploads/doc/12.Control_bark_eating_caterpillar.pdf

www.nematology.ucdavis.edu/faculty/westerdahl/courses/nemas/meloidogyneincognita.htm

www.tnau.ac.in/eagri/eagri50/ENTO331/lecture29/lec029.pdf, accessed on 14-07-2014

Yadava, C. P. S. 1969. Combating fruit sucking moths. *Indian Horticulture*, **13**(3): 32-34.

Zaher, M. A. and Yousef, A. A. 1972. Biology of the False Spider Mite *Tenuipalpus punicae* P. & B. in U. A. R. (Acarina — Tenuipalpidae). *Journal of Applied Entomology*, **70**(1-4): 23–29.